Kinetic Studies in GeO$_2$/Ge System

Frontiers in Semiconductor Technology

Semiconductor technology has been perhaps the most prominent technology industry in modern society over the past seventy years. Facing the future, emerging technologies are constantly shaping the industry and promoting its continuous development.

Outstanding young scientists from various technology sectors have been invited to join this book series. Through this platform, the aim is for the books within the series to provide new insights and contributions to the development of modern semiconductor technology. The scope of the series is wide, covering semiconductor physics, materials, device processes, equipment, and IC design methods, amid many other topics, while studies involving case studies and applied settings will also be prominent. The titles included in the series are designed to appeal to students, researchers, and professionals across semiconductor science and engineering, as well as interdisciplinary researchers from many scientific disciplines.

Please contact us if you have an idea for a book for the series.

Titles in the series currently include:

Kinetic Studies in GeO$_2$/Ge System
A Retrospective from 2021
Sheng-Kai Wang

Kinetic Studies in GeO$_2$/Ge System

A Retrospective from 2021

Sheng-Kai Wang

CRC Press
Taylor & Francis Group
Boca Raton   London

CRC Press is an imprint of the
Taylor & Francis Group, an **informa** business

This book is published with financial support from a grant-in-aid from the Youth Innovation Promotion Association of the Chinese Academy of Sciences.

First edition published 2022
by CRC Press
6000 Broken Sound Parkway NW, Suite 300, Boca Raton, FL 33487–2742

and by CRC Press
4 Park Square, Milton Park, Abingdon, Oxon, OX14 4RN

CRC Press is an imprint of Taylor & Francis Group, LLC

© 2022 Sheng-Kai Wang

Library of Congress Cataloging-in-Publication Data
Names: Wang, Shengkai, 1984– author.
Title: Kinetic studies in GeO₂/Ge system : a retrospective from 2021 / Sheng-Kai Wang.
Description: First edition. | Boca Raton, FL : CRC Press, 2022. | Series: Semiconductor technology | Revision of author's thesis (Ph. D.)—University of Tokyo, 2011. | Includes bibliographical references.
Identifiers: LCCN 2021055725 (print) | LCCN 2021055726 (ebook) | ISBN 9781032257440 (hbk) | ISBN 9781032258485 (pbk) | ISBN 9781003284802 (ebk)
Subjects: LCSH: Metal oxide semiconductors, Complementary—Materials. | Germanium—Electric properties.
Classification: LCC TK7871.99.M44 W3575 2022 (print) | LCC TK7871.99.M44 (ebook) | DDC 621.39/732—dc23/eng/20211227
LC record available at https://lccn.loc.gov/2021055725
LC ebook record available at https://lccn.loc.gov/2021055726

ISBN: 978-1-032-25744-0 (hbk)
ISBN: 978-1-032-25848-5 (pbk)
ISBN: 978-1-003-28480-2 (ebk)

DOI: 10.1201/9781003284802

Typeset in Minion
by Apex CoVantage, LLC

Contents

Figures

Tables

Tables

Preface

THIS BOOK IS MAINLY BASED ON MY DOCTORAL DISSERTATION IN THE Department of Materials Engineering, University of Tokyo from 2008 to 2011. During the period of doctoral study, I followed Professor Akira Toriumi to learn the reaction kinetics in the GeO_2/Ge system. Compared with my dissertation, there are some updates in this book, including: 1) research progress achieved after I joined the Institute of Microelectronics of the Chinese Academy of Sciences; 2) new results with new data or figures; and 3) retrospection on my current opinion in 2021 in terms of "Author's Note".

During 2008–2011, Ge was regarded as a promising candidate to replace Si for advanced complementary metal oxide semiconductor (CMOS) devices because of its high mobility. At present (2021), the opinion towards the possibility of using pure Ge as a candidate has already changed. However, as the technology node goes from 22nm to 5nm and beyond, Ge content in the channel is increasing. Therefore, no matter whether pure Ge CMOS could be used or not, studying the most fundamental GeO_2/Ge stack still maintains its value. The kinetic study on GeO_2/Ge is still insufficient to meet the requirement for the control of GeO_2/Ge stacks. This book investigates the reaction kinetics in GeO desorption, GeO_2 crystallization, and Ge oxidation in the GeO_2/Ge system, aiming to provide the fundamental idea about the GeO_2/Ge interface and give an answer as to why Ge behaves so differently from Si in principle.

Many people have made valuable suggestions for this book. I am particularly grateful to Professor Akira Toriumi and all my professors, colleagues, and students at the University of Tokyo and the Institute of Microelectronics of the Chinese Academy of Sciences. The completion of this book could not have been done without their support. And this book is also supported by a grant-in-aid from the Youth Innovation Promotion Association of the Chinese Academy of Sciences.

Sheng-Kai Wang
Beijing

CHAPTER 1

Introduction

1.1 SCALING TRENDS OF MOS TECHNOLOGY

The tremendous development and wide spread of information technology (IT) have brought enormous influence on our society in the aspects of traditional manufacturing, business, management, and human lifestyle. Nowadays, the IT industry is the most important industry in most developed countries. As is well known to all, the information revolution benefits from the invention of various types of computers and telecommunication equipment. And all these computers and telecommunication equipment are based on microelectronic devices.

At present, the key component of the microelectronic industry is the metal oxide semiconductor (MOS) integrate circuit (IC). Generally, the development of MOS technology indicates the developing level of microelectronic technology. The performance of an integrated circuit is improved by reducing the dimensions (scaling) of the MOS devices, which increases the devices' number per integrated circuit and decreases the cost. In 1964, Intel co-founder Gordon E. Moore forecast the rapid pace of technology innovation. He stated that transistor density on integrated circuits doubles about every two years. This prediction is popularly known as "Moore's Law".[1, 2] As is shown in Figure 1.1, which lists the development trend of IC complexity,[3a, b] "Moore's Law" has continued well for more than half a century. For logic applications, by simply reducing the feature size of the silicon MOS field effect transistor (FET), the device packing density and performance have been significantly improved while power consumption

DOI: 10.1201/9781003284802-1

1

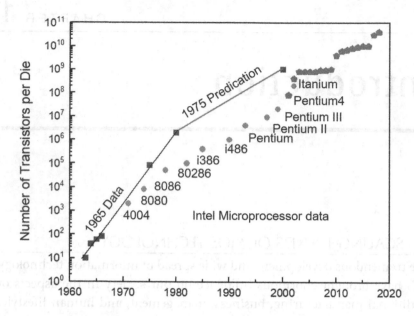

FIGURE 1.1   Illustration of Moore's law in 2010 and 2020.[3a, b]

and cost are reduced. And the IC industry successfully entered the 22nm era in 2011.[4a] (Author's note: In 2020, the most advanced technology is 5nm for TSMC; 10 billion transistors can be put into one 5nm chip. In this foreground, the next complete process node is 3nm after 5nm, and TSMC is also in research and development. Compared with the first generation 5nm process, the performance of the chip will be improved by 10% to 15%, the energy consumption will be reduced by 25% to 30%, and the density of transistors will be increased by 70%. Mass production of 3nm will be done in 2022, as announced by the TSMC on World Semiconductor Conference on August 26, 2020).[4b]

1.1.1 Scaling Rule

Equivalent scaling-down was first proposed by R. H. Dennard in 1974.[5] His basic principle is to improve IC performance by scaling down the device size while the electric field in MOSFETs is kept constant. This scaling rule is called the constant-field (CE) law. When the device size, source voltage, and threshold voltage are scaled with a factor κ, the resulting device parameters are as listed in Table 1.1.

TABLE 1.1 Equivalent Scaling Rules

Physical Parameter	CE Law	CV Law	QCE Law
Device Size (L, W, t_{ox})	$1/\kappa$	$1/\kappa$	$1/\kappa$
Source Voltage	$1/\kappa$	1	λ/κ
Impurity Concentration	κ	κ^2	$\lambda\kappa$
Threshold Voltage	$1/\kappa$	1	λ/κ
Source-Drain Current	$1/\kappa$	κ	λ^2/κ
Electric Field	1	κ	λ
Gate Delay	$1/\kappa$	$1/\kappa^2$	$1/\lambda\kappa$
Power Consumption	$1/\kappa^2$	κ	λ^3/κ^2
Power Density	1	κ^3	λ^3
Packaging Density	κ^2	κ^2	κ^2

Remarkable success has been achieved by applying CE law in equivalent scaling; however, there are still several main problems to be overcome in CE law:

1. Too small V_T scaling is harmful to the capability of IC against disturbance.

2. The width of the depletion region at the source/drain is impossible to equivalently scale down.

3. Frequently changing the source voltage (V_{DD}) is not convenience for actual application.

Therefore, to overcome the problems induced by the CE law, constant-voltage (CV) law has been proposed to keep the source voltage V_{DD} and threshold voltage V_T constant. However, CV law also causes some other negative impacts on IC performance and reliability, such as the heat sink problem. Generally, CV law is not applicable for highly scaled (<1μm) devices.

Eventually, by taking both the power density problem in the highly scaled region and keeping the V_T not too small, the actual solution does not follow an equivalent way. Generally, if the device size is scaled by $1/\kappa$, the source voltage is scaled by λ/κ. This scaling rule is called the quasi-constant-field (QCE) rule. The parameters for CV law and the QCE rule are also listed in Table 1.1.

1.1.2 Challenge in Advanced Technology Node

The scaling of planar bulk silicon (Si) channel MOSFETs has been found to be successful in the past 30 years. However, when looking back on the innovations of MOSFETs in the past 15–20 years, we found the technology for device scaling is really innovative. Figure 1.2 shows the representative technology for device scaling in the past and forthcoming technology nodes.

In the 130nm node, traditional bulk Si MOSFETs technology is demonstrated to be very successful. When entering sub 100nm node, the limitations of traditional bulk Si MOSFETs are found to be more and more severe. Therefore, diverse new materials and technology, such as strained-Silicon (90–65–45),[7, 8] high-k/metal gate (45–32),[9, 10] FinFET (32–22–16 and beyond),[11] Ge/III-V channel (16 and beyond),[12, 13] nanowire (16 and beyond),[14] and/or grapheme/nanotubes (16 and beyond)[15–17] are in use and/or potential candidates for future application. (Author's note: Ge/III-V channel, grapheme/nanotubes seem to have been eliminated from the scope of major companies; they still maintain important value for some niche markets, such as silicon photonics, high frequency circuits, anti-radiation, etc.).

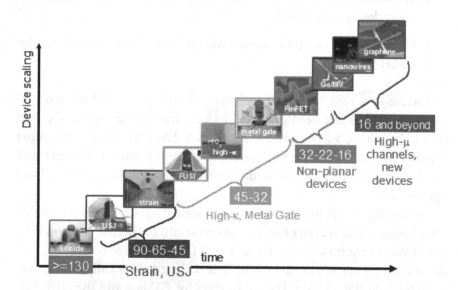

FIGURE 1.2 Representative technology upon device dimension shrinkage.[6]

The saturation of Si MOSFETs drain current I_{DS} upon dimension shrinkage may limit the prospect of future scaling in the beyond-22nm node. The drive current of MOSFETs per gate width can be given by

$$I_{ON} \approx qv_s N_s^{source} \qquad (1\text{-}1)$$

where q is the element charge, v_s is the carrier velocity near the source edge, and N_s^{source} is the surface carrier concentration near the source edge.[18, 19] While N_s^{source} is simply determined by V_g and C_g, v_s is dependent on the channel length, as schematically shown in Figure 1.3. In long-channel MOSFETs, the carrier transport is dominated by a diffusive transport, where I_{ON} is determined by the product of the electric field near the source edge E_s and low field mobility near the source edge u_s, meaning that low field carrier mobility is an important parameter in determining I_{ON}, as shown in Figure 1.3(a). This is the common picture of carrier conduction in MOSFETs under stationary carrier transport. As long as carrier scattering events in the channels occur sufficiently often and the stationary transport dominates the carrier transport, this model basically holds, even under the existence of velocity saturation, where E_s is affected

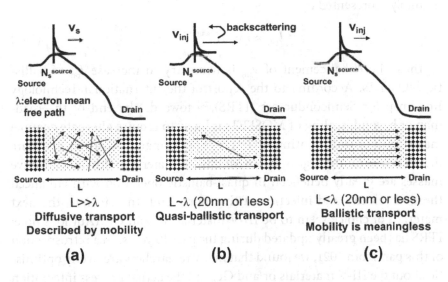

FIGURE 1.3 Schematic diagrams of carrier-transport models to determine on current in MOSFET. (a) Conventional transport model. (b) Quasi-ballistic transport model. (c) Full-ballistic transport model.

by velocity saturation, and thus, the u_s dependence of I_{ON} becomes significantly weaker than the linear one.

On the other hand, as the channel length becomes shorter, non-stationary transport becomes more dominant, where sufficient numbers of scattering events do not occur inside the channels. This situation, as shown in Figure 1.3(b), has been formulated as quasi-ballistic transport by Lundstrom et al.[18] and has been quantitatively analyzed in detail by many research groups.

$$I_{ON} \approx q v_{inj} \frac{1-r}{1+r} N_s^{source} \qquad (1-2)$$

where v_{inj} is injection velocity at the top of the barrier near the source edge, and r is the backscattering rate near the source region. Since r is directly related to u_s, the enhancement of mobility can be still important in increasing I_{on} under the quasi-ballistic transport regime.[20, 21]

Furthermore, when channel length becomes much shorter, probably down to less than 10nm in Si MOSFETs,[22] and no carrier-scattering events occur inside the channel, the carrier transport is dominated by full ballistic transportation, as shown in Figure 1.3(c). Here, I_{on} in MOSFETs under this ballistic transport, which has also been formulated by Natori,[23] is simply represented by

$$I_{ON} \approx q v_{inj} N_s^{source} \qquad (1-3)$$

Thus, the enhancement of v_{inj} is necessary to increase I_{on} of ballistic MOSFETs. According to the report of the International Technology Roadmap for Semiconductors (ITRS),[24] towards the end of the roadmap or beyond, scaling of MOSFETs is likely to require alternate channel materials in order to continue to improve performance. To attain adequate drive current for the highly scaled MOSFETs, materials with light effective masses are greatly beneficial in quasi-ballistic operation with enhanced thermal velocity and injection at the source end. In principle, the next materials of choice seem to be III-V materials or/and Ge. (Author's note: ITRS has been greatly updated during the past 10 years. In a retrospection of this part from 2021, we found that many researchers are overly optimistic about the III-V materials or/and Ge, and the actual process integration of such novel channel materials into the conventional Si line is far more difficult than expected.)

1.2 MOTIVATION OF EMERGING GE AS CHANNEL MATERIAL

Both Ge and III-V materials are promising candidates to replace Si for future MOSFETs because of their high electron and/or hole mobility. Table 1.2 lists the bulk mobility of Si, Ge, and several typical III-V materials.

As shown in Table 1.2, most III-V materials are promising candidates to replace Si as channel materials for N-MOS. However, in general, the hole mobility does not show a symmetric behavior. Although InSb shows excellent behavior on both electron and hole mobility, its narrow band gap limits the possibility of making complementary MOS (CMOS). Generally, narrow band gap is beneficial for lower-voltage operation, but the leakage current, the performance at high operation temperature, is not satisfied. Moreover, lack of suitable p-channel material in III-V compounds is another drawback. Although it is argued that use of III-V materials for N-MOS and Ge for P-MOS may be a better solution, it requires the co-integration of Ge and III-V materials on the Si substrate. However, this process inevitably increases the cost, process complexity, and yield risk. For III-V materials, besides channel selection, there are many issues and challenges in applying III-V compounds in highly scale nodes,[26] including crystal growth with controllable defects and stress,[27-29] high-k dielectric/III-V stack formation,[30, 31] surface passivation,[32-35] co-integration of Ge and III-V materials, compromise of activation temperature between Ge and III-V compounds,[36] source/drain (S/D) formation, etc.

Compared to Si, Ge is now viewed as one of the most promising channel materials because of high electron and hole mobility, process compatibility with Si MOS technologies, lower temperature process due to a lower melting point than Si, and possible voltage scaling due to a narrow

TABLE 1.2   Band Gap and Mobility in Semiconductors[25]

	Si	Ge	GaAs	InSb	GaSb	InP
Band Gap E_g at 300K (eV)	1.12	0.66	1.42	0.17	0.72	1.35
Hole Mobility μ_h (cm²/Vs)	450	1900	400	1250	850	150
Electron mobility μ_e (cm²/Vs)	1500	3900	8500	800000	5000	4600

band gap. Ge shows nearly 4 times faster hole mobility and 2.5 times faster electron mobility. Well-balanced electron and hole mobility enables the use of Ge to make complementary MOS (CMOS). Moreover, on the basis of the as-developed SiGe technology for strained silicon, pure Ge epitaxy on Si is not expected to be an obstacle. Therefore, if only a single material is used for a highly scaled MOSFETs channel, Ge will be a promising choice. In the emerging research materials (ERM) critical assessment in ITRS,[26] all alternate channel materials (Ge, III-V, graphene, nanotube, nanowires) were viewed to have potentially better mobility than silicon CMOS. However, from an integration perspective, all the options were considered less capable than CMOS (a score of 2), but the Ge alternate channel got a higher score (~1.8) than III-Vs (~1.6), nanowires (~1.6), carbon nanotubes (~1.4), and graphene (~1.4). (Author's note: The author was once involved in the National Science and Technology Major Project of China (also known as 02 Project, 2011–2016). Four institutions were working together in a project titled Integration Technology for Silicon-Based High-Mobility Materials and Novel Devices to develop technologies for next-generation ICs. Researchers at Tsinghua University are focusing on the integration and realization of high-quality SiGe or germanium-on-silicon, while a team from the Institute of Semiconductors of CAS is grappling with how to integrate III-Vs with silicon; the Peking University team is concentrating on the memristor development using HfO$_x$-based resistive random-access memory (RRAM), and the team from the Institute of Microelectronics of CAS is working on process integration technology, including the high-κ stack engineering and process integration technology development. Although some interesting papers and patents were published and issued at that time, due to the high risk for the major foundries, the related work is still limited to the fundamental research level even up to now.[37])

1.3 CHALLENGES IN THE GE-CHANNEL PROCESS

1.3.1 Historical Study in Ge

In retrospect, over the past 70 years, Ge has had a great history, with two Nobel Prize, the world's first transistor, and the world's first IC.[38, 39] On December 16, 1947, the first transistor action was experimentally observed by John Bardeen and Walter Brattain in Bell Labs. They placed two close gold-plated probe tips onto poly-crystalline Ge to build the first point-contact transistor.[40] In 1956, Bardeen and Brattain, together with

Shockley, were awarded the Nobel Prize "for their researches on semiconductors and their discovery of the transistor effect".[38]

In mid-1958, Jack Kilby concentrated on the problem in circuit design
that was commonly called the "tyranny of numbers" and finally came
to the conclusion that manufacturing the circuit components en masse
in a single piece of semiconductor material could provide a solution. On
September 12, 1958, Jack Kilby presented his findings to the management:
he showed them a piece of germanium with an oscilloscope attached and
pressed a switch; the oscilloscope showed a continuous sine wave, proving
that his IC worked and thus that he solved the problem. This was the first
IC in the world. [41]

In 2000, Kilby was awarded the Nobel Prize in Physics "for basic work
in information and communication technology and for his part in the
invention of the integrated circuit".[39]

The first transistor and IC were made of Ge; however, discrete transistors were transitioned from Ge to Si in 1960s. The reasons are mainly as
follows:

1. Ge is more expensive than Si because the element is less abundant.
   The earth's abundance of Ge is estimated to be 1.8 ppm,[42] while the
   earth's abundance of Si is 277,100 ppm.[43] The major raw material for
   Si wafer fabrication is sand, and there is lots of sand (SiO_2) available.

2. Ge has a smaller band gap and thus becomes intrinsic at relatively
   lower temperatures. We have a good story to illustrate why the
   transition from Ge to Si was necessary in the 1950s: Motorola was
   a company that made car radios and one of the first companies to
   make transistor radios. In the 1950s, the transistors were made of
   Ge. However, if a car was parked in the sun on a really hot sum-
   mer day, the radio no longer worked. Why? Because Ge had become
   intrinsic; this caused the n-type and p-type regions to lose their distinct properties, and, as a result, the bipolar junction transistors no
   longer worked. This gave Motorola a strong motivation to replace Ge
   bipolar transistors with Si bipolar transistors.

3. It is difficult to form an excellent high-k dielectric/Ge or GeO_2/Ge
   interface comparable to Si/SiO_2. The thermal instability of the dielec-
   tric/Ge system strongly hampers the development of Ge-MOS technology and limits its application.

In order to apply Ge channel as a new performance booster for beyond 22nm node, the cost problem could be alleviated by using Ge epitaxy technology. High-temperature conditions can be properly controlled by reducing the power density together with the improved the heat-sink system. Therefore, how to form a good high-k/Ge interface becomes a key issue. (Author's note: This idea is still right even in 2021; Ge epitaxy is still the best solution for Ge/Si hetero-integration. And high-k/Ge interface is still a key issue, especially in the direction of reliability.)

1.3.2 GeO₂/Ge Stack Degradation

Ge is a promising candidate to replace Si in the future beyond scaling devices because of its high mobility. However, unlike Si, it is difficult to grow an insulating oxide comparable to SiO_2/Si on Ge. The lack of thermodynamic stability at the high-k/Ge interface hampers the development of Ge metal-oxide-semiconductor (MOS) devices due to desorption of GeO.[44, 45] Generally, GeO_2/Ge has been considered the most fundamental MIS interface in Ge, as SiO_2/Si is in Si MOSFETs. Our previous study showed that below 700°C, GeO desorption was not derived from the decomposition of GeO_2 itself but from its reaction with Ge substrate.[46] For GeO_2/Ge, GeO desorbs at around 550°C, whereas no GeO desorption can be observed from a GeO_2/SiO_2/Si structure. Desorption of GeO from a high-k/Ge system deteriorates the surface and interface quality,[47] which finally leads to the electrical characteristics degradation.[47, 48] Therefore, suppressing the GeO desorption is essential to improve the interface quality so as to fabricate the Ge-based devices for future applications.

1.3.3 Ge-Passivation Techniques

To crack the hard nut of this technique's challenge, many attempts, such as plasma oxidation, ozone or O radical oxidation, high-pressure oxidation (HPO), surface or interfacial nitridation, S-passivation, F-passivation, Cl-passivation, Si-, Sr-, Be- interfacial insertion, and rare-earth introduction have been applied to improve the quality of high-k/Ge or oxide/Ge system stability.[49–59] This subsection will give a review of passivation methods.

1. Plasma oxidation.

An electron cyclotron resonance (ECR) system with a divergent magnetic field is able to generate a high-density ($\sim 10^{12}$ cm⁻³) and well-energy-collimated (10–30 eV) plasma stream even in a low-pressure atmosphere of $10^{-2} \sim 10^{-1}$Pa. Moreover, since the same amount of negative and positive charges of electrons and ions irradiates the substrate surface, the ECR plasma stream causes less plasma-induced damage, even to the insulator surfaces.

Fukuda et al. have reported very successful results on the application of ECR to form a GeO_2/Ge interface.[60] Fukuda et al. have demonstrated that the Al_2O_3/GeO_2/Ge interface prepared by ECR plasma irradiation exhibits a good C-V characteristic with very low D_{it} of 6×10^{10} cm^{-2}eV^{-1}. (Author's note: Since Fukuda's work, plasma oxidation has been successfully developed by the Takagi group from the University of Tokyo to a higher level to meet the requirements of low D_{it} and EOT at the same time. And various encouraging results about high mobility Ge-MOSFETs have been demonstrated by R. Zhang et al. from 2011–2017.[61])

2. Ozone or O radical treatment.

The main advantages of ozone or radical oxidation[62–64] are 1) low process temperature, 2) surface orientation-independent oxidation rate (not for ozone), 3) less temperature dependence of oxidation rate, 4) less interface roughness after oxidation, and 5) low electrically active interface defects. Therefore, ozone or radical oxidation is considered one of the key processes in the future MOSFET technology with a high-mobility channel material. Damage from high-temperature annealing is a common problem as it also degrades interfaces between germanium and high-κ dielectrics. To prevent this from occurring, we have inserted a high-quality interfacial GeO_x layer between the high-κ dielectric and the germanium substrate. We introduce this with a cycling ozone oxidation method, which involves repeatedly depositing Al_2O_3 via atomic layer deposition and then performing *in situ* ozone oxidation at about 300°C. The transistors that result have a very low density of interface traps and capacitance-voltage characteristics that are dispersion free, as shown in Figure 1.4.[50]

Kobayashi et al. have performed the radical oxidation on a Ge substrate by the slot-plane-antenna (SPA) with high-density radical oxidation and investigated the physical/electrical properties of grown GeO_2 for the gate dielectric interfacial layer applications.

By the SPA radical oxidation, no substrate orientation dependence of growth rate attributed to highly reactive oxygen radicals with low

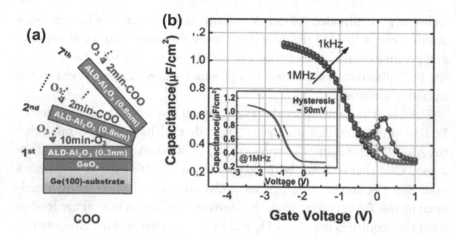

FIGURE 1.4 (a) Schematic of fabrication procedure of Ge MOS gate stacks for samples with cycling ozone oxidation (COO) treatments. (b) C-V characteristics for COO sample, where the inset shows the C-V hysteresis. D_{it} of 1.9×10^{11} cm^{-2} eV^{-1} is obtained by inserting GeO$_x$ passivation layer with COO treatment.

Source: Reproduced from X. Yang et al., "Al₂O₃/GeOₓ gate stack on germanium substrate fabricated by in situ cycling ozone oxidation method", *Appl. Phys. Lett.* 105, 092101 (2014), with the permission of AIP Publishing.

oxidation activation energy has been demonstrated, which is highly beneficial to three-dimensional structure devices, such as multi-gate field-effect transistors, to form conformal gate dielectrics. The electrical properties of an aluminum oxide Al₂O₃ metal-oxide-semiconductor gate stack with a GeO₂ interfacial layer were investigated, showing very low interface state density $D_{it} \sim 1.4 \times 10^{11}$ cm^{-2} eV^{-1} [51].

3. High-pressure oxidation (HPO)

From the view point of thermodynamic control, increasing the O₂ pressure is considered a possible solution to suppress GeO desorption so as to form a high-quality GeO₂ film.[49, 65] Lee et al. have carried out oxidation process in 70 atm O₂ ambient at 550°C, followed by low temperature annealing (LOA) at 400°C in 1 atm O₂ ambient for a long time (over 30 minutes). It has been demonstrated that the GeO₂/Ge interface prepared by HPO exhibits perfect capacitance-voltage (C-V) characteristic with almost no hysteresis and frequency dispersion. And D_{it} less than 10^{11}

$cm^{-2}eV^{-1}$ has been achieved. Details about the HPO system will be discussed in chapter 2.

4. Surface and/or interface nitridation.

In order to obtain a more stable native passivation layer, either thermal or plasma anodic nitridation was applied to form Ge oxynitrides (GeO_xN_y),[66, 67] which were, however, not very scalable. Instead, the use of high-k dielectric directly[68, 69] or high-k on oxynitride[70] on Ge has shown some promise. However, a passivating GeO_xN_y layer prior to HfO_2 to obtain functional capacitor and transistor devices does not provide sufficient passivation of the Ge interface.[71] Dimoulas et al. had already carried out experiment on HfO_2/GeON/Ge structure (GeON was prepared from a remote RF plasma source), and they found that GeO_x or GeON interfacial layers are unstable and are partly dissociated by reaction with HfO_2, releasing Ge or oxidized Ge complexes inside the HfO_2 layer. As a result, the current-voltage characteristics are non-ideal, exhibiting large hysteresis and frequency dispersion and high interface density of states D_{it}.[72–74] Sugawara et al.[75] had studied the water solubility of GeO_xN_y layer prepared by remote inductively coupled plasma (ICP) or radial line slot antenna (RLSA), respectively. Their results showed that high-pressure remote plasma (radical dominant plasma) nitridation forms water soluble nitrogen-germanium bonding, and low pressure RLSA plasma (ion dominant plasma) nitridation forms water-resistant nitrogen-germanium bonding. Although the hydrogen mixture and ion dominant processes increase nitrogen concentration, those processes accompany water-insoluble sub-oxide formation (ICP) and plasma charging damage (RLSA), respectively.

Another possible solution to Ge surface passivation is the direct nitridation of Ge because the lack of Ge-O bonds in pure nitrides essentially eliminates the formation of Ge sub-oxides and desorption of GeO. Maeda et al. have succeeded in forming pure Ge_3N_4/Ge MIS structures.[52, 76, 77]

There are two important factors in the formation of pure Ge_3N_4. One is to use reactive nitrogen species because a clean Ge surface without any incorporation of oxygen is totally inert to NH_3.[78] They employed atomic nitrogen radicals generated by plasma nitrogen sources. Another important factor is the realization of oxygen- and carbon-free clean Ge surface just before nitridation.[79] In order to realize this requirement, they employed thermal cleaning at 600°C in a UHV chamber; thus, the oxygen and carbon could be eliminated by thermal desorption. After that, the Ge surface received successive plasma-enhanced nitridation and formed a 3nm thick interlayer (grown at 100°C for 30 minutes). This is the only report of the formation of pure Ge$_3$N$_4$ so far. According to their wet etching experiment using HF and ultra-pure water (UPW), Ge$_3$N$_4$ films are water insoluble but soluble in HF.

Although surface nitridation of Ge (i.e., GeON or Ge$_3$N$_4$) is thermally more stable than GeO$_2$, the existence of nitrogen at the interface possibly degrades carrier mobility due to increased oxide defects and interface states, which act as Coulomb scattering centers in a similar manner to directly attached SiON and/or SiN in Si MOS.[80]

5. Si passivation.

Passivation of a Ge surface with silicon covering is also an attractive technique because the Si/Ge epitaxial interface is maintained, which is less flawed than the Ge-insulator interface. In most case, Si-passivation was used for PMOS.[55, 81–84] According to N. Taoka et al.,[84] the improvement of inversion layer hole mobility is attributed to the reduction in interface charges and the separation of mobile carriers by increasing the Si passivation layer thickness. This approach, however, has a trade-off between device performance and scalability of gate dielectric. The drawback of the passivation by Si is that a low-interfacial SiO$_x$ layer can be inevitable, which may prevent a dielectric film from effective oxide thickness scaling. In addition, the Si conduction band is lower than the Ge conduction band,[85] so electrons will be mostly populated in the Si layer with lower electron mobility in the inversion layer of Ge NMOS.

6. Sr interlayer insertion.

The degradation of the GeO$_2$/Ge stack causes a large gate leakage current. To overcome this problem, Kamata et al.[56, 86] have proposed inserting a strontium germanide (SrGe$_x$) interlayer to a high-k/Ge interface to exclude the Ge oxide and demonstrated a significant J$_g$ reduction and both Ge p-FET and Ge n-FET with the same gate stack (LaAlO$_3$/SrGe$_x$/Ge).[87] Under a thermal budget of 420°C in the back-end off-line (BEOL) process, equivalent oxide thickness (EOT) scaling down of the gate stack with $SrGe_x$ interlayer to sub-1nm has been achieved.

In Ge MOSFETs, V_{th} shifts towards positive bias direction independent of channel type (p-FET and n-FET) with an increase in sub-threshold slope are a common issue. It is generally believed that Fermi level pinning (FLP) near the charge neutrality level (CNL) is the reason for the V_{th} shifts, which is caused by the increase in D_{it} around CNL.[86, 88] The interfacial insertion of a $SrGe_x$ layer is considered to play a role in passivating both donor type traps and acceptor type ones near the CNL, leading to small sub-threshold slope (SS) and suppressing V_{th} shift.

7. Sulfur passivation.

Sulfur (S) passivation by aqueous ammonium sulfide $((NH_4)_2S)$ treatment is reported to have lower interface-state density than NH_3-nitridation-passivated samples.[54] Moreover, it has been demonstrated that S passivation can reduce the EOT of gate stack, whereas it can maintain the gate leakage current.[89]

Furthermore, S passivation can improve the thermal stability of the Ge/high-k gate stack. Thus, S passivation improves the thermal stability of Ge gate stack. The improved interface properties in S-treated samples can possibly be explained by the GeOS interfacial layer, which may suppress the Ge out diffusion due to fewer Ge-O bonds and provide more stable interface properties, whereas for the samples without S passivation, the surface consists of GeO_x, and it may enhance Ge out diffusion[90] and result in poor interface property.

8. Fluorine passivation.

Although hydrogen (H) passivation by forming gas annealing (FGA) is effective to terminate the dangling bonds at SiO_2/Si interface, it has been reported that FGA might not be an effective way for Ge.[91, 92] Fluorine (F) passivation is considered an alternative for Ge passivation because Ge-F has higher bonding energy (~5.04 eV) than Ge-H bond (~3.34 eV), and it has been shown that F can segregate near high-k/Ge interface and reduce both interface states and high-k bulk traps.[53] Xie et al.[93] performed F passivation by CF_4 plasma treatment to GeO_2 passivated HfO_2-gated Ge

MOS capacitors in an inductively coupled plasma chamber. PDA at 500°C for 30 seconds was then performed for all samples. After TaN metal electrode formation, FGA at 350°C for 1 hour was performed for some samples. Compared to devices with only FGA, devices with F incorporation exhibit about 12% and 17% higher peak and high-field hole mobilities, respectively. This is ascribed to the better interface quality after the F passivation. The reduction in D_{it} suggests that F passivation can be an effective method for Ge interface after gate dielectric formation to passivate interface defects.

9. Rare-earth introduction.

Various transition metal high-κ dielectrics such as HfO_2 and ZrO_2 have been tried on Ge to find a suitable gate dielectric for CMOS applications.[68, 94] Although abrupt interface between Ge and high-κ dielectrics looks quite good in TEM images in terms of surface roughness, the electrical characteristic at the high-k/Ge interface is negative. The resulting D_{it} values at these high-k/Ge interfaces are always very high.

Compared to the transition metal high-k group, $LaLuO_3$ attracts much attention because of its high crystallization temperature (>800°C), high k value (>20), and good stability on Ge.[57, 58] Recently, Tabata et al. reported excellent electrical characteristic by using $LaLuO_3$ as a gate dielectric on Ge.[57] The ideal C-V curves indicate that $LaLuO_3$ is a promising candidate for high-k dielectric selection on Ge. On the basis of a thermal desorption spectroscopy (TDS), it has also been demonstrated that $LaLuO_3$ is more effective in suppressing the GeO desorption than HfO_2. Therefore, capping the GeO₂/Ge stack with some rare-earth oxide like $LaLuO_3$ may open a new direction for Ge passivation.

(Author's note: New methods on Ge-MOS interface passivation using crystalline dielectric, such as BeO, ZrO_2, HfO_2, etc., have been developed in the past decade and have been regarded as a promising direction for both maintaining low D_{it} and EOT. For example, epitaxial growth of BeO on Ge and III-V was demonstrated in 2011 and 2012. By applying the atomic ALD method, epitaxial BeO has been demonstrated to be a good choice for Ge and III-V passivation because of its advantageous properties in the lattice (domain) match, large band gap (10.6 eV), thermal stability, and thermal conductivity.[95, 96])

1.4 OBJECTIVE

On the one hand, although Ge MOSFETs' electrical characteristics have been significantly improved by applying the aforementioned techniques, without sufficient knowledge of the desorption mechanism of GeO from a GeO_2/Ge stack, the process optimization and technology evolution remain limited.

On the other hand, compared to the well-known SiO_2/Si system, systematic investigation of the GeO_2/Ge system will be helpful in revealing the intrinsic difference between the GeO_2/Ge system and the SiO_2/Si system.

To realize the quality control of the GeO_2 dielectric and the GeO_2/Ge interface, the objective of this study could be summarized as follows:

1. Reveal the reasons for the quality degradation of the GeO_2/Ge system and demonstrate the intrinsic difference between the GeO_2/Ge system and the SiO_2/Si system.

2. Provide the guideline for GeO_2 dielectric quality and GeO_2/Ge interface quality control on the basis of knowledge of the GeO_2/Ge system.

1.5 BOOK ORGANIZATION

In this subchapter, the outline of this book will be given in summary.

Chapter 1 reviews the development history of MOS technology and discusses the motivation of developing Ge as channel material for beyond 22nm node. It also reviews the challenges of Ge as a channel material, including the introduction of the history of Ge study, a brief discussion of the GeO_2/Ge degradation, and a brief review of Ge passivation techniques. In the final part of this chapter, the objectives of this study and thesis organization are introduced.

Chapter 2 describes the fabrication and thermal desorption theory and characterization methods used throughout this study. Detailed fabrication flows of sample preparation and the basic principles behind all fabrication equipment are discussed. The thermal desorption spectroscopy setups and desorption theory used in this study are also introduced.

Chapter 3 mainly focuses on the desorption kinetics of GeO from GeO_2/Ge. After a brief introduction to the historical desorption study in the $Ge:O_2/Ge$ system, the mechanism of GeO desorption from GeO_2/Ge

is systematically revealed by using Ge and O isotope-tracing experiments, together with thermal desorption spectroscopy study. Moreover, desorption of GeO in the nonuniform region is investigated from the viewpoint of a morphology study. A kinetic model in the uniform desorption region is proposed in this chapter. In addition, the relationship between GeO desorption and electrical degradation in GeO_2/Ge is investigated.

Chapter 4 focuses on the GeO_2 crystallization behavior and studies the relationship between GeO_2 crystallization and GeO_2/Ge interface reaction and establishes the connection between GeO nonuniform desorption and GeO_2 crystallization. A unified model integrating uniform/nonuniform GeO desorption, GeO_2 crystallization, and GeO_2/Ge interface redox reaction is proposed.

Chapter 5 discusses the oxidation kinetics of Ge, especially in the direction of active oxidation. Moreover, guidelines for the control of the GeO_2/Ge system are presented and discussed. Finally, a fundamental consideration of the difference between GeO_2/Ge and SiO_2/Si is presented.

REFERENCES

1. G. E. Moore, "Cramming more components onto integrated circuits", *Electronics*, **38** (1965) 114.
2. Although originally calculated as a doubling every year, Moore later refined the period to two years, see G. E. Moore, "Progress in Digital Integrated Electronics", *Tech. Dig. IEEE Int. Electron Devices Meeting*, 1975, p. 11.
3. a. ftp://download.intel.com/research/silicon/Gordon_Moore_ISSCC_021003.pdf; b. www.futuretimeline.net/data-trends/9.htm
4. a. P. Otellini, Intel Developer Forum 2009, "Paul Otellini just held up a 22nm wafer at his Intel Developer Forum keynote, saying that chips with the technology would be out in the second half of 2011"; b. Zhenqiu Luo, World Semiconductor Conference, Aug. 26, 2020, Nanjing.
5. R. H. Dennard, F. H. Gaensslen, H. N. Yu, V. L. Ideout, E. Bassous, and A. R. LeBlanc, "Design of ion-implanted MOSFET's with very small physical dimensions", *IEEE. J Solid-State Circuits*, **SC-9** (1974) 256.
6. Courtesy of M. Houssa, KU Leuven.
7. M. L. Lee, E. A. Fitzgerald, M. T. Bulsara, M. T. Currie, and A. Lochtefeld, "Strained Si, SiGe, and Ge channels for high-mobility metal-oxide-semiconductor field-effect transistors", *J. Appl. Phys.*, **97** (2005) 011101.
8. S. E. Thompson, M. Armstrong, C. Auth, S. Cea, R. Chau, G. Glass, T. Hoffman, J. Klaus, Z. Y. Ma, B. Mcintyre, A. Murthy, B. Obradovic, L. Shifren, S. Sivakumar, S. Tyagi, T. Ghani, K. Mistry, M. Bohr, and Y. El-Mansy, "A logic nanotechnology featuring strained-silicon", *IEEE Electron Dev. Lett.*, **25** (2004) 191.
9. J. Robertson, "High dielectric constant gate oxides for metal oxide Si transistors", *Rep. Prog. Phys.*, **69** (2006) 327.

10. R. Chau, S. Datta, M. Doczy, B. Doyle, J. Kavalieros, and M. Metz, "High-kappa/metal-gate stack and its MOSFET characteristics", *IEEE Electron Dev. Lett.*, **25** (2004) 408.

11. C. E. Smith, H. Adhikari, S.-H. Lee, B. Coss, S. Parthasarathy, C. Young, B. Sassman, M. Cruz, C. Hobbs, P. Majhi, P. D. Kirsch, and R. Jammy, "Dual channel FinFETs as a single high-k/metal gate solution beyond 22nm node", *Tech. Dig., IEEE Int. Electron Devices Meeting*, 2009, p. 309.

12. C. H. Lee, T. Nishimura, T. Tabata, S. K. Wang, K. Nagashio, K. Kita, and A. Toriumi, "Ge MOSFETs performance: Impact of Ge interface passivation", *Tech. Dig., IEEE Int. Electron Devices Meeting*, 2010, p. 416.

13. R. J. R. Hill, J. Huang, J. Barnett, P. Kirsch, and R. Jammy, "III-V MOSFETs: Beyond silicon technology", *Solid State Tech.*, **53** (2010) 17.

14. M. Luisier, "Phonon-limited and effective low-field mobility in n- and p-type [100]-, [110]-, and [111]-oriented Si nanowire transistors", *Appl. Phys. Lett.*, **98** (2011) 032111.

15. J. S. Moon, D. Curtis, S. Bui, T. Marshall, D. Wheeler, I. Valles, S. Kim, E. Wang, X. Weng, and M. Fanton, "Top-gated graphene field-effect transistors using graphene on Si (111) wafers", *IEEE Electron Dev. Lett.*, **31** (2010) 1193.

16. Q. Zhang, Y. Q. Lu, H. G. Xing, S. J. Koester, and S. O. Koswatta, "Scalability of atomic-thin-body (ATB) transistors based on graphene nanoribbons", *IEEE Electron Dev. Lett.*, **31** (2010) 531.

17. T. Durkop T, B. M. Kim, and M. S. Fuhrer, "Properties and applications of high-mobility semiconducting nanotubes", *J. Phys.: Condens. Matter*, **16** (2004) R553.

18. M. Lundstrom and Z. Ren, "Essential physics of carrier transport in nanoscale MOSFETs", *IEEE Trans. Electron Devices*, **49** (2002) 133.

19. M. Lundstrom and J. Guo, *Nanoscale Transistors*, New York: Springer-Verlag, 2006.

20. A. Lochtefeld and D. A. Antoniadis, "On experimental determination of carrier velocity in deeply scaled NMOS: How close to the thermal limit?", *IEEE Electron Device Lett.*, **22** (2001) 95.

21. R. Ohba and T. Mizuno, "Nonstationary electron/hole transport in sub-0.1 um MOS devices: Correlation with mobility and low-power CMOS application", *IEEE Trans. Electron Devices*, **48** (2001) 338.

22. H. Tsuchiya, K. Fujii, T. Mori, and T. Miyoshi, "A quantumcorrected Monte Carlo study on quasi-ballistic transport in nanoscale MOSFETs", *IEEE Trans. Electron Devices*, **53** (2006) 2965.

23. K. Natori, "Ballistic metal—oxide—semiconductor field effect transistor", *J. Appl. Phys.*, **76** (1994) 4879.

24. The International Technology Roadmap for Semiconductors, *Process Integration, Devices, and Structure*, 2009, 2009 ed., p. 4.

25. S. M. Sze, *Physics of Semiconductor Devices*, New York: Wiley, 1981, 2nd ed., p. 789.

26. The International Technology Roadmap for Semiconductors, *Emerging Research Materials*, Washington, Semiconductor Industry Association, 2009, 2009 ed., pp. 2–6.

27. H. Tanoto, S. F. Yoon, W. K. Loke, K. P. Chen, E. A. Fitzgerald, C. Dohrman, and B. Narayanan. "Heteroepitaxial growth of GaAs on (100) Ge/Si using migration enhanced epitaxy", *J. Appl. Phys.*, **103** (2008) 104901.

28. K. Chilukuri, M. J. Mori, C. L. Dohrman, and E. A. Fitzgerald. "Monolithic CMOS-compatible AlGaInP visible LED arrays on silicon on lattice-engineered substrates (SOLES)", *Semicond. Sci. Technol.*, **22** (2007) 29.

29. T. Nishinaga, "Microchannel epitaxy: An overview", *J. Cryst. Growth*, **237–239** (2002) 1410.

30. J. F. Zheng, W. Tsai, M. Hong, J. Kwo, and T. P. Ma. "Ga$_2$O$_3$(Gd$_2$O$_3$)/Si$_3$N$_4$ dual-layer gate dielectric for InGaAs enhancement mode MOSFET with channel inversion", *Appl. Phys. Lett.*, **91** (2007) 223502.

31. N. Goel, P. Majhi, W. Tsai, M. Warusawithana, D. G. Schlom, M. B. Santos, J. S. Harris, and Y. Nishi, "High-indium content InGaAs MOS capacitor with amorphous LaAlO$_3$ gate dielectric", *Appl. Phys. Lett.*, **91** (2007) 093509.

32. P. T. Chen, Y. Sun, E. Kim, P. C. McIntyre, W. Tsai, M. Garner, P. Pianetta, Y. Nishi, and C. O. Chui, "HfO$_2$ gate dielectric on (NH$_4$) S passivated (100) GaAs grown by ALD", *J. Appl. Phys.*, **103** (2008) 034106.

33. B. Shin, J. Cagnon, R. D. Long, P. K. Hurley, S. Stemmer, and P. C. McIntyre., "Unpinned interface between Al$_2$O$_3$ gate dielectric layer grown by atomic layer deposition and chemically treated n-In$_{0.53}$Ga$_{0.47}$As(001)", *Electrochem. Solid-State Lett.*, **12** (2009) G40.

34. S. Oktyabrsky, V. Tokranov, S. Koveshnikov, M. Yakimov, R. Kambhampati, H. Bakhru, R. Moore, and W. Tsai, "Interface properties of MBE-grown MOS structures with InGaAs/InAlAs buried channel and in-situ high-k oxide", *J. Cryst. Growth*, **311** (2009) 1950.

35. I. J. Ok, H. Kim, M. Zhang, F. Zhu, S. Park, J. Yum, H. Zhao, D. Garcia, P. Majhi, N. Goel, W. Tsai, C. K. Gaspe, M. B. Santos, and J. C. Lee, "Self-aligned n-channel MOSFET on high-indium-content In$_{0.53}$Ga$_{0.47}$As and InP using PVD HfO$_2$ and silicon passivation layer", *Appl. Phys Lett.*, **92** (2008) 202903.

36. E. Simoen, A. Satta, A. D'Amore, T. Janssens, T. Clarysse, K. Martens, B. DeJaeger, A. Benedetti, I. Hoflijk, B. Brijs, M. Meuris, and W. Vandervorst, "Ion implantation issues in the formation of shallow junctions in germanium", *Mater. Sci. in Semicon. Proc*, **9** (2006) 634.

37. S. K. Wang, "Turbo charging the channel", *Compound Semiconductor*, **1–2** (2016) 36.

38. http://nobelprize.org/nobel_prizes/physics/laureates/1956/index.html

39. http://nobelprize.org/nobel_prizes/physics/laureates/2000/index.html

40. http://en.wikipedia.org/wiki/File:Replica-of-first-transistor.jpg

41. www.ti.com/corp/graphics/press/image/on_line/co1034.jpg

42. Kenneth Barbalace, Periodic Table of Elements—Silicon—Si. Environmental Chemistry.com. 1995–2011.

43. Kenneth Barbalace, Periodic Table of Elements—Germanium—Ge. Environmental- Chemistry.com. 1995–2011.

44. K. Prabhakaran and T. Ogino, "Oxidation of Ge (100) and Ge (111) surfaces: An UPS and XPS study", *Surf. Sci.*, **325** (1995) 263.

45. J. Oh and J. C. Campbell, "Thermal desorption of Ge native oxides and the loss of Ge from the surface", *J. Electron. Mater.*, **33** (2004) 364.
46. K. Kita, S. Suzuki, H. Nomura, T. Takahashi, T. Nishimura, and A. Toriumi, "Direct evidence of GeO volatilization from GeO_2/Ge and impact of its suppression on GeO_2/Ge metal-insulator-semiconductor characteristics", *Jpn. J. Appl. Phys.*, **47** (2008) 2349.
47. Y. Kamata, "High-k/Ge MOSFETs for future nanoelectronics", *Materials Today*, **11** (2008) 30.
48. H. Seo, F. Bellenger, K. B. Chung, M. Houssa, M. Meuris, M. Heyns, and G. Lucovsky, "Extrinsic interface formation of HfO_2 and Al_2O_3/GeO_x gate stacks on Ge (100) substrates", *J. Appl. Phys.*, **106** (2009) 044909.
49. C. H. Lee, T. Tabata, T. Nishimura, K. Nagashio, K. Kita, and A. Toriumi, "Ge/GeO_2 interface control with high-pressure oxidation for improving electrical characteristics", *Appl. Phys. Express*, **2** (2009) 071404.
50. X. Yang, S. K. Wang, X. Zhang, B. Sun, W. Zhao, H. D. Chang, Z. H. Zeng, and H. G. Liu, "Al_2O_3/GeO_x gate stack on germanium substrate fabricated by in situ cycling ozone oxidation method", *Appl. Phys. Lett.*, **105** (2014) 092101.
51. M. Kobayashi, G. Thareja, M. Ishibashi, Y. Sun, P. Griffin, P. Pianetta, J. McVittie, K. Saraswat, and Y. Nishi, "Radical oxidation of germanium for interface gate dielectric GeO_2 formation in metal-insulator-semiconductor gate stack", *J. Appl. Phys.*, **106** (2009) 104117;
52. T. Maeda, T. Yasuda, M. Nishizawa, N. Miyata, Y. Morita, and S. Takagi, "Pure germanium nitride formation by atomic nitrogen radicals for application to Ge metal-insulator-semiconductor structures", *J. Appl. Phys.*, **100** (2006) 014101.
53. K. Saraswat, C. O. Chui, T. Krishnamohan, D. Kim, A. Nayfeh, and A. Pethe, "High performance germanium MOSFETs", *Mater. Sci. Eng. B*, **135** (2006) 242.
54. R. Xie, W. He, M. Yu, and C. Zhu, "Effects of fluorine incorporation and forming gas annealing on high-k gated germanium metal-oxide-semiconductor with GeO_2 surface passivation", *Appl. Phys. Lett.*, **93** (2008) 073504.
55. M. Frank, S. J. Koester, M. Copel, J. A. Ott, V. K. Paruchuri, H. Shang, and R. Loesing, "Hafnium oxide gate dielectrics on sulfur-passivated germanium", *Appl. Phys. Lett.*, **89** (2006) 112905.
56. J. Mitard, C. Shea, B. DeJaeger, A. Pristera, G. Wang, M. Houssa, G. Eneman, G. Hellings, W-E. Wang, J. C. Lin, F. E. Leys, R. Loo, G. Winderickx, E. Vrancken, A. Stesmans, K. DeMeyer, M. Caymax, L. Pantisano, M. Meuris, and M. Heyns, "Impact of EOT scaling down to 0.85nm on 70nm Ge-pFETs technology with STI", *Tech. Dig. VLSI Symp.*, 2009, p. 82.
57. Y. Kamata, A. Takashima, Y. Kamimuta, and T. Tezuka, "New approach to form EOT-scalable gate stack with strontium germanide interlayer for high-k/Ge MISFETs", *Tech. Dig. VLSI Symp.*, 2009, p. 78.
58. T. Tabata, C. H. Lee, K. Kita, and A. Toriumi, "Direct $LaLuO_3$/Ge gate stack formation by interface scavenging and subsequent low temperature O_2 annealing", *ECS Trans.*, **33**(3) (2010) 375.

59. J. J. Gu, Y. Q. Liu, M. Xu, G. K. Celler, R. G. Gordon, and P. D. Ye, "High performance atomic-layer-deposited $LaLuO_3$/Ge-on-insulator p-channel metal-oxide-semiconductor field-effect transistor with thermally grown GeO_2 as interfacial passivation layer", *Appl. Phys. Lett.*, **97** (2010) 012106.

60. Y. Fukuda, Y. Yazaki, Y. Otani, T. Sato, H. Toyota, and T. Ono, "Low-temperature formation of high-quality GeO_2 interlayer for high-κ gate dielectrics/Ge by electron-cyclotron-resonance plasma techniques", *IEEE Trans. Electron Devices*, **57** (2010) 282.

61. R. Zhang, T. Iwasaki, N. Taoka, M. Takenaka, and S. Takagi, "High mobility Ge pMOSFETs with ~1 nm thin EOT using Al_2O_3/GeO_x/Ge gate stacks fabricated by plasma post oxidation", *IEEE VLSI Symposia -Technology (VLSI)*, 2011.6.14–2011.6.16; R. Zhang, T. Iwasaki, N. Taoka, M. Takenaka, and S. Takagi, "High-mobility Ge pMOSFET with 1-nm EOT Al_2O_3/GeO_x/Ge gate stack fabricated by plasma post oxidation", *IEEE Trans. Electron Devices*, **2** (2012) 335; R. Zhang, T. Iwasaki, N. Taoka, M. Takenaka, and S. Takagi, "Impact of channel orientation on electrical properties of Ge p- and n-MOSFETs with 1-nm EOT Al_2O_3/GeO_x/Ge gate-stacks fabricated by plasma postoxidation", *IEEE Trans. Electron Devices*, **11** (2014) 3668;

62. T. Ohmi, S. Sugawa, K. Kotani, M. Hirayama, and A. Morimoto, "New paradigm of silicon technology", *Proc. IEEE*, **89** (2001) 394.

63. K. Sekine, Y. Saito, M. Hirayama, and T. Ohmi, "Highly reliable ultrathin silicon oxide film formation at low temperature by oxygen radical generated in high-density krypton plasma", *IEEE Trans. Electron Devices*, **48** (2001) 1550.

64. T. Sugawara, S. Matsuyama, M. Sasaki, T. Nakanishi, S. Murakawa, J. Katsuki, S. Ozaki, Y. Tada, T. Ohta, and N. Yamamoto, "Characterization of ultra thin oxynitride formed by radical nitridation with slot plane antenna plasma", *Jpn. J. Appl. Phys., Part I*, **44** (2005) 1232.

65. O. Knacke, O. Kubaschewski, and K. Hesselmann, *Thermodynamical Properties of Inorganic Substances*, Berlin: Springer-Verlag, Part I, 1991, p. 767.

66. O. J. Gregory, E. E. Crisman, L. Pruitt, D. J. Hymes, and J. J. Rosenberg, "Electrical characterization of some native insulators on germanium", *Proc. Mater. Res. Soc. Symp.*, **76** (1987) 307.

67. Z. Sun and C. Liu, "Plasma anodic oxidation and nitridation of Ge(111) surface", *Semicond. Sci. Technol.*, **8** (1993) 1779.

68. C. O. Chui, S. Ramanathan, B. B. Triplett, P. C. McIntyre, and K. C. Saraswat, "Germanium MOS capacitors incorporating ultrathin high-k gate dielectric", *IEEE Electron Device Lett.*, **23** (2002) 473.

69. H. Kim, C. O. Chui, K. C. Saraswat, and P. C. McIntyre, "Local epitaxial growth of ZrO_2 on Ge (100) substrates by atomic layer epitaxy", *Appl. Phys. Lett.*, **3** (2003) 2647.

70. W. P. Bai, N. Lu, J. Liu, A. Ramirez, D. L. Kwon, D. Wristers, A. Ritenour, L. Lee, and D. Antoniadis, "Ge MOS characteristics with CVD HfO2 gate dielectrics and TaN gate electrode", *VLSI Symp. Tech. Dig.* 2003, p. 121.

71. A. Dimoulas, D. P. Brunco, S. Ferrari, J. W. Seo, and M. M. Heyns, "Interface engineering for Ge metal-oxide-semiconductor devices", *Thin Solid Films*, **515** (2007) 6337.

72. A. Dimoulas, G. Mavrou, G. Vellianities, E. K. Evanelou, N. Boukos, M. Houssa, and M. Caymax, "HfO2 high-k gate dielectrics on Ge (100) by atomic oxygen beam deposition", *Appl. Phys. Lett.,* **86** (2005) 032908.

73. A. Dimoulas, G. Mavrou, G. Vellianities, E. K. Evangelou, A. Sotiropoulos, "Intrinsic carrier effects in HfO$_2$-Ge metal-insulator-semiconductor capacitors", *Appl. Phys. Lett.,* **86** (2005) 223507.

74. A. Dimoulas, in: E. Gusev (Ed.), *Defects in High-k Gate Dielectric Stacks,* NATO Science Scries II, vol. 220, Netherlands: Springer, 2006, p. 237.

75. T. Sugawara, R. Sreenivasan, and P. C. McIntyre, "Mechanism of germanium plasma nitridation", *J. Vac. Sci. Technol. B,* **24** (2006) 2442.

76. T. Maeda, T. Yasuda, M. Nishizawa, N. Miyata, Y. Morita, and S. Takagi, "Ge metal-insulator-semiconductor structures with Ge$_3$N$_4$ dielectrics by direct nitridation of Ge substrates", *Appl. Phys. Lett.,* **85** (2004) 3181.

77. T. Maeda, M. Nishizawa, Y. Morita, S. Takagi, "Role of germanium nitride interfacial layers in HfO$_2$/germanium nitride/germanium metal-insulator-semiconductor structures", *Appl. Phys. Lett.,* **90** (2007) 072911.

78. D. Aubel, M. Diani, J. L. Bischoff, D. Bolmont, and L. Kubler, "Strict thermal nitridation selectivity between Si and Ge used as a chemical probe of the outermost layer of Si$_{1-x}$Ge$_x$ alloys and Ge/Si(001) or Si/Ge(001) heterostructures", *J. Vac. Sci. Technol. B,* **12** (1994) 2699.

79. S. Takagi, T. Maeda, N. Taoka, M. Nishizawa, Y. Morita, K. Ikeda, Y. Yamashita, M. Nishikawa, H. Kumagai, R. Nakane, S. Sugahara, and N. Sugiyama, "Gate dielectric formation and MIS interface characterization on Ge", *Microelectronic Eng.,* **84** (2007) 2314.

80. M. A. Schmidt, F. L. Terry, B. P. Mathur, and S. D. Senturia, "Inversion layer mobility of MOSFETs with nitrided oxide gate dielectrics", *IEEE Trans Electron Devices,* **35** (1988) 1627.

81. Y. Wang, Y. Z. Hu, E. A. Irene, "In-situ investigation of the passivation of Si and Ge by electron cyclotron resonance plasma enhanced chemical vapor deposition of SiO$_2$", *J. Vac. Sci. Technol. B,* **14** (1996) 1687.

82. G. G. Fountain, R. A. Rudder, S. V. Hattangady, D. J. Vitkavage, R. J. Markunas, J. B. Posthill, "Electrical and microstructural characterization of ultrathin Si interlayer used in silicon dioxide/germanium-based MIS structure", *Electron. Lett.,* **24** (1988) 692.

83. D. J. Vitkavage, G. G. Fountain, R. A. Rudder, S. V. Hattangady, and R. J. Markunas, "Gating of germanium surfaces using pseudomorphic silicon interlayers", *Appl. Phys. Lett.,* **53** (1988) 692.

84. N. Taoka, M. Harada, Y. Yamashita, T. Yamamoto, N. Sugiyama, Shin-ichi Takagi, "Effects of Si passivation on Ge metal-insulator-semiconductor interface properties and inversion hole mobility", *Appl. Phys. Lett.,* **92** (2008) 113511.

85. C. G. Van de Walle and R. M. Martin, "Theoretical calculations of heterojunction discontinuities in the Si/Ge system", *Phys. Rev. B*, **34** (1986) 5621.
86. Y. Kamata, A. Takashima, Y. Kamimuta, and T. Tezuka, "High-k/Ge p- & n-MISFETs with Strontium Germanide Interlayer for EOT Scalable CMIS Application", *Tech. Dig. VLSI Symp.*, 2010, p. 211.
87. J. Huang, P. D. Kirsch, J. Oh, S. H. Lee et al., "Mechanisms limiting EOT scaling and gate leakage currents of high-k/metal gate stacks directly on SiGe and a method to enable sub-1nm EOT", *Tech. Dig. VLSI Symp.*, 2008, p. 82.
88. D. Kuzum, T. Krishnamohan, A. Nainani, Y. Sun, P. A. Pianetta, H. S-. P. Wong, and K. C. Saraswat, "Experimental demonstration of high mobility Ge NMOS", *Tech. Dig., IEEE Int. Electron Devices Meeting*, 2009, p. 453.
89. R. Xie and C. Zhu, "Effects of sulfur passivation on germanium MOS capacitors with HfON gate dielectric", *IEEE Electron Dev. Lett.*, **28** (2007) 976.
90. N. Lu, W. Bai, A. Ramirez, C. Mouli, A. Ritenour, M. L. Lee, D. Antoniadis, and D. L. Kwong, "Ge diffusion in Ge metal oxide semiconductor with chemical vapor deposition HfO$_2$ dielectric", *Appl. Phys. Lett.*, **87** (2005) 051922.
91. V. V. Afanas'ev, Y. G. Fedorenko, and A. Stesmans, "Interface traps and dangling-bond defects in (100)Ge/HfO$_2$", *Appl. Phys. Lett.*, **87** (2005) 032107.
92. J. R. Weber, A. Janotti, P. Rinke, and C. G. Van de Walle, "Dangling bond defects and hydrogen passivation in germanium", *Appl. Phys. Lett.*, **91** (2007) 142101.
93. R. L. Xie, T. H. Phung, W. He, M. B. Yu, and C. X. Zhu, Interface-engineered high-mobility high-K/Ge p-MOSFETs with 1nm equivalent oxide thickness", *IEEE Trans. Elec. Devices*, **56** (2009) 1330.
94. E. P. Gusev, H. Shang, M. Copel, M. Grilbeyuk, C. D'Emic, P. Kozlowski, and T. Zabel, "Microstructure and thermal stability of HfO$_2$ gate dielectric deposited on Ge(100)", *App. Phys. Lett.*, **85** (2004) 2334.
95. J. H. Yum, T. Akyol, M. Lei, D. A. Ferrer, Todd. W. Hudnall, M. Downer, C. W. Bielawski, G. Bersuker, J. C. Lee, and S. K. Banerjee, "Inversion type InP metal oxide semiconductor field effect transistor using novel atomic layer deposited BeO gate dielectric", *App. Phys. Lett.*, **99** (2011) 033502.
96. S. K. Wang, B.-Q. Xue, B. Sun, H.-L. Liang, Z.-X. Mei, W. Zhao, X.-L. Du, and H.-G. Liu, "Comprehensive study of interface passivation in Ge-MOSFETs -Control the interfacial layer for high performance devices-", IEEE 11th International Conference on Solid-State and Integrated Circuit Technology, 29 Oct.–1 Nov. 2012, Xi'an, China, 2012.

CHAPTER 2

Fabrication and Characterization Methods

2.1 INTRODUCTION

This chapter will first introduce the sample preparation of GeO_2/Ge stacks and metal-insulator-semiconductor (MIS) capacitor. The detailed sample setups and processing steps will be explained. The second part concentrates on thermal desorption spectroscopy (TDS) to present a brief introduction of TDS and give a theoretical review of the desorption process.

2.2 STACK PREPARATION

2.2.1 Ge Wafer Cleaning

The Ge wafer used in this study is four-inch Czochralski (CZ) crystal wafer made by Umicore. Figure 2.1 shows a picture of the Ge (100) wafer.

The as-received Ge wafer is easily oxidized with the formation of a thin native oxide layer. Different from GeO_2, this native oxide contains a certain amount of sub-oxide (Ge^{2+}). Due to the low quality of this sub-oxide, the as-received Ge wafer, without cleaning treatment, is not suitable for making devices.

To fully remove the native oxide and other containments, the Umicore Czochralski (CZ) crystal wafers are cleaned through a-four-step process:

DOI: 10.1201/9781003284802-2

25

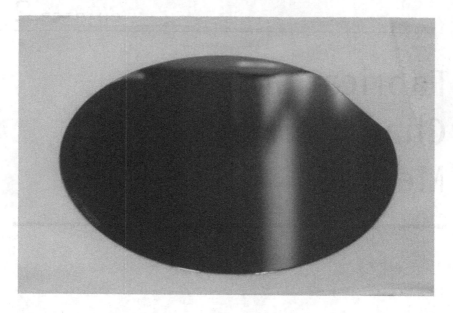

FIGURE 2.1   4-inch Ge (100) wafer used in this study.

1. The wafers (usually 1cm × 1cm) are cleaned in methanol using an ultrasonic cleaner (100 kHz) for ten minutes, followed by a de-ionized water (18.3M ohm-cm DIW) cleaning. The purpose of this process is to remove the adsorbed organic molecules and Ge particles.

2. The wafers are then rinsed in a 25% HCl solution for one minute, followed by DIW cleaning to remove possible metal contaminants.

3. The wafers are rinsed in an H_2O_2/ammonia/H_2O (1:0.5:100) solution, followed by DIW cleaning to neutralize the residual HCl and form a protective GeO_2 layer.

4. They are then rinsed in a 5% HF solution for three minutes, followed by DIW rinsing for 15 seconds. Ge wafers are dried with a pure nitrogen blow immediately after that. This process can etch the GeO_2 formed by the former process.

As a result, Ge with almost no surface natural oxides is obtained. Figure 2.2 shows the x-ray photoelectron spectroscopy (XPS) spectrum of an as-cleaned Ge wafer. For comparison, a Ge wafer undergoing just ultrasonic DIW cleaning is also shown as a control.

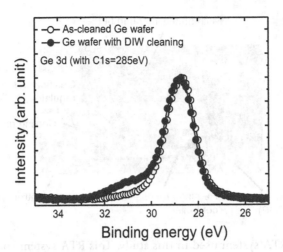

FIGURE 2.2   XPS Ge 3d spectra of as-cleaned Ge wafer (black open dot) and Ge wafer with only DIW ultrasonic cleaning (blue closed dot). After the four-step cleaning process, sub-oxide is totally removed.

2.2.2 Film Formation and Annealing Treatment

In this subsection, methods for film formation will be introduced.

1. Rapid thermal annealing (RTA) treatment.[1]

Thermal annealing processes are often used in modern semiconductor fabrication for defects recovery, oxide growth, lattice recovery, or impurity electrical activation of doped or ion-implanted wafers.[2] For the sputtered film, RTA treatment is also performed to eliminate the damage to the film from the sputtering process. In this study, the RTA system is used for thermal oxide growth and post-deposition annealing of sputtered GeO_2 and other high-k materials. Figure 2.3 shows a picture of our rapid thermal annealing system, in which the QHC-P610CP RTA system from the ULVAC Company is used.

Prior to every annealing cycle, existing gases inside the anneal chamber and gas lines were flowed with high-grade pure N_2 (99.99995%), followed by pumping and purging to minimize the possibility of contamination by other existing gases or particles. This RTA system is heated up by an infrared lamp heating furnace. The gas flow rate is kept at 250sccm (standard cubic centimeters per minute) for annealing

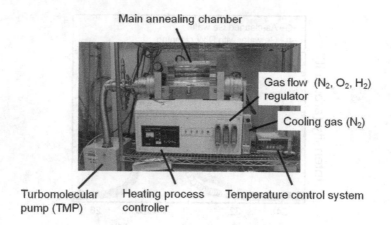

FIGURE 2.3 RTA system used in this study. This RTA system contains several modules, including a main chamber, a gas flow regulator, an infrared lamp heating system, a cooling gas system, and a turbomolecular pump.

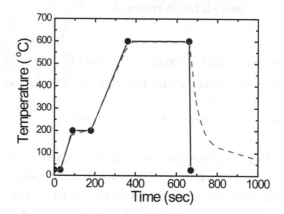

FIGURE 2.4 Typical program for annealing treatment in this RTA system. First, the temperature is increased from room temperature (RT) to 200°C to remove the possible water adsorbant and make the system ready for a rapid heating process. After that, the temperature is rapidly heated up to a target temperature (600°C in this example). After the annealing treatment, N$_2$ flow is usually added at a rate of 2L/min to enhance the cooling process. And the sample will not be taken out until it reaches 80°C. Red dashed line is the real temperature.

cycles. A two-inch Si (100) wafer is used as a sample holder. All annealed samples are removed from the chamber at below 80°C. Figure 2.4 shows a typical (600°C as a target temperature) program for annealing treatment in this RTA system.

2. Sputtering deposition.[3]

Sputter deposition is a physical vapor deposition (PVD) method of depositing thin films by sputtering a block of source material (target) onto a substrate. Sputtered atoms ejected into the gas phase are not in their thermodynamic equilibrium state and tend to deposit on all surfaces in the vacuum chamber. A substrate placed in the chamber will be coated with a thin film as shown in Figure 2.5.

Sputtering sources are usually magnetrons that utilize strong electric and magnetic fields to trap electrons close to the surface of the magnetron, which is known as the target. The electrons follow helical paths around the magnetic field lines, undergoing more ionizing collisions with gaseous neutrals near the target surface than would otherwise occur. The sputter gas is inert, typically argon. The extra argon ions created as a result of these collisions lead to a higher deposition rate. It also means that the plasma can be sustained at a lower pressure. The sputtered atoms are neutrally charged and so are unaffected by the magnetic trap. Charge build-up on insulating targets can be avoided with the use of RF sputtering, in which the sign of the anode-cathode bias is varied at a high rate. RF sputtering works well to produce highly insulating oxide films but only with the added expense of RF power supplies and impedance matching networks.

One important advantage of sputtering as a deposition technique is that the deposited films have the same composition as the source material.

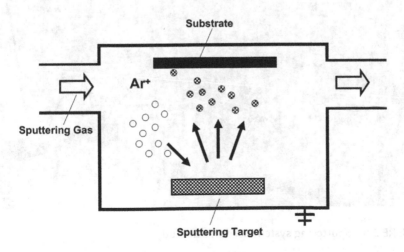

FIGURE 2.5   Illustration of the principle of sputtering.

The equality of the film and target stoichiometry might be surprising since the sputter yield depends on the atomic weight of the atoms in the target. One might therefore expect one component of an alloy or mixture to sputter faster than the other components, leading to an enrichment of that component in the deposit. Sputter deposition also has an advantage over molecular beam epitaxial (MBE) due to its speed. The higher rate of deposition results in lower impurity incorporation because fewer impurities are able to reach the surface of the substrate in the same amount of time. Sputtering methods are consequently able to use process gases with far higher impurity concentrations than the vacuum pressure that MBE methods can tolerate. During sputter deposition, the substrate may be bombarded by energetic ions and neutral atoms. Ions can be deflected with a substrate bias and neutral bombardment can be minimized by off-axis sputtering, but only at a cost in deposition rate. Plastic substrates cannot tolerate the bombardment and are usually coated via evaporation.

In this study, we used the RF sputtering method, and the argon gas flow rate was 22sccm. The sputter power is changed from 20–100w in order to tune the deposition rate. Figure 2.6 shows the sputtering system we used in this study.

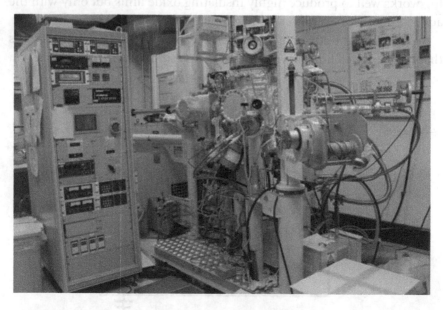

FIGURE 2.6   Sputtering system used in this study.

For GeO$_2$ film deposition, using only argon gas usually results in an oxygen-deficient GeO$_2$ because the deposition rates for Ge and O atoms are different. To overcome this problem, adding additional O$_2$ flow during the deposition process will be helpful. In our study, we added 0.6sccm O$_2$ flow together with 22sccm Ar flow during the GeO$_2$ deposition. Although the quality of sputtered GeO$_2$ may differ from the thermal grown ones, concerning the thermal desorption experiments in this study, we believe the quality of sputtered GeO$_2$ is reliable for thermal desorption study and the desorption behavior of GeO from sputtered GeO$_2$, and the thermal grown case are quite similar (also see Figure 3.5).

(Author's note: The sputtering system in Toriumi Lab is a miracle. It was made by ULVAC. Over 30 IEDM/VLSI papers were published by using this system, making me realize the advantage of PVD for gate dielectric engineering and oxide physics study. In general, a sputtering system is more stable than a CVD system; moreover, it is very easy to form composite oxides with the co-sputtering process. Furthermore, for most gate oxides, the sputtering targets can be easily bought in a lower cost than the CVD precursor.)

3. High-pressure oxidation (HPO)[4]

Figures 2.7(a) and (b) show a schematic illustration and a camera illustration of the HPO system used for high-quality GeO$_2$ formation.

It is mainly composed of a high-pressure oxygen cylinder, a pressure regulator, several valves, a rotary vacuum pump, and a tube furnace that consists of a quartz oxidation tube enclosed in a steel pressure vessel. The HPO system is evacuated to approximately 1Pa by rotary pump after the cleaned Ge wafers are placed into the quartz oxidation tube. After that, the furnace chamber surrounding the steel pressure vessel is heated to a thermal oxidation temperature. Temperature calibration of the HPO furnace was carried out in the temperature range of 200°C to 600°C for precise measurements.

(Author's note: The HPO system in Toriumi Lab is actually very cheap but effective in forming high quality GeO$_2$. Totally speaking, its development includes three generations at least. The success of the HPO system strongly indicates that putting an "expensive" idea into a "cheap" system is one of the best ways to start a new direction. Don't worry too much about

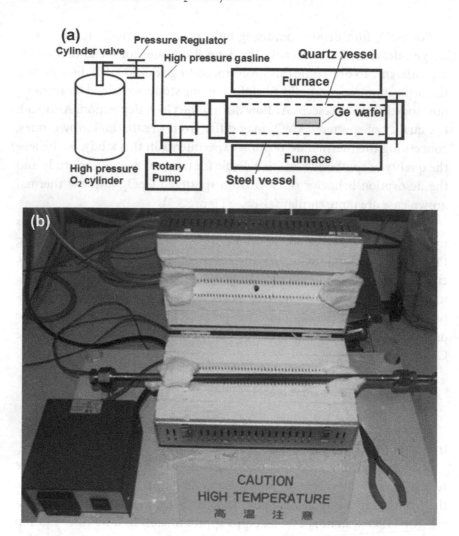

FIGURE 2.7   (a) Schematic illustration of HPO system. (b) HPO system used in this study.

the quality of the system; making it run and do some basic demonstration is important and meaningful.)

4. Electron beam physical vapor deposition (EBPVD)

EBPVD[5, 6] is a form of physical vapor deposition in which a target anode is bombarded with an electron beam given off by a charged tungsten filament under high vacuum. The electron beam causes atoms from the target to transform into the gaseous phase. These atoms then precipitate into solid form, coating everything in the vacuum chamber (within line of sight) with a thin layer of the anode material.

In an EBPVD system, the deposition chamber is evacuated to a pressure of 10^{-9} Torr. The material to be evaporated is in the form of ingots. Electron beams can be generated by thermionic emission, field electron emission, or the anodic arc method. The generated electron beam is accelerated to a high kinetic energy and focused towards the ingot. By increasing the applied power, most of the kinetic energy of the electrons is converted into thermal energy as the beam bombards the surface of the ingot. The surface temperature of the ingot increases resulting in the formation of a liquid melt. Although some of the incident electron energy is lost in the excitation of x-rays and secondary emission, the liquid ingot material evaporates under vacuum.

The ingot itself is enclosed in a copper crucible, which is cooled by water circulation. The level of molten liquid pool on the surface of the ingot is kept constant by vertical displacement of the ingot. The number of ingot feeders depends upon the material to be deposited. The schematic and camera photo of our EBPVD system is shown in Figure 2.8(a) and (b).

5. Vacuum evaporation by heating.

Vacuum evaporation (including sublimation) is a PVD process in which material from a thermal vaporization source reaches the substrate without collision with gas molecules in the space between the source and substrate. The trajectory of the vaporized material is "line-of-sight". In our system, vacuum evaporation takes place in a gas pressure range of 10^{-7} to 10^{-9} Torr. For an appreciable deposition rate to be attained, the material vaporized must reach a temperature at which its vapor pressure is 10m Torr or higher. Typical vaporization sources are resistively heated stranded wires, boats, or crucibles (for vaporization temperatures below 1500°C). Figures 2.9(a) and (b) show the schematic and camera photo of our vacuum evaporation system.

2.2.3 Electrode Deposition

Resistive thermal evaporation is one of the most commonly used metal deposition techniques. It consists of vaporizing a solid material (pure

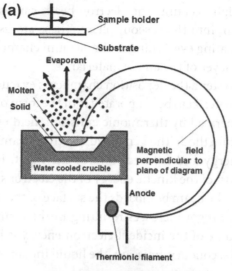

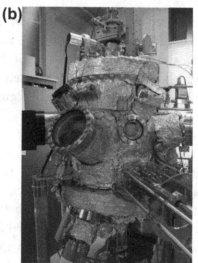

FIGURE 2.8   (a) Schematic of EBPVD system. (b) EBPVD system used in this study.

metal, eutectic, or compound) by heating it to sufficiently high temperatures and recondensing it onto a cooler substrate to form a thin film. As the name implies, the heating is carried out by passing a large current through a filament container that has a finite electrical resistance. The choice of this filament material is dictated by the evaporation temperature

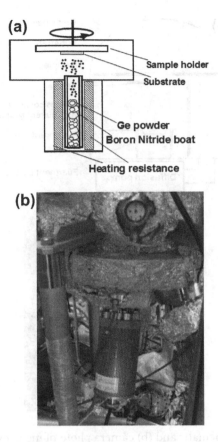

FIGURE 2.9   (a) Schematic and (b) camera photo of vacuum evaporation system.

and its inertness to alloying/chemical reaction with the evaporant. This technique is also known as "indirect" thermal evaporation since a supporting material is used to hold the evaporant.

Once the metal has evaporated, its vapor undergoes collisions with the surrounding gas molecules inside the evaporation chamber. As a result, a fraction are scattered within a given distance during their transfer through the ambient gas. The mean free path for air at 25°C is approximately 45 and 4500cm at pressures of 10^{-4} and 10^{-6} Torr, respectively. Therefore, pressures lower than 10^{-5} Torr are necessary to ensure a straight-line path for most of the evaporated species and for a substrate-to-source distance of approximately 10 to 50cm in a vacuum chamber. A good vacuum is also a prerequisite for producing contamination-free deposits. A scheme of the metal evaporation system used in this study is showed in Figure 2.10.

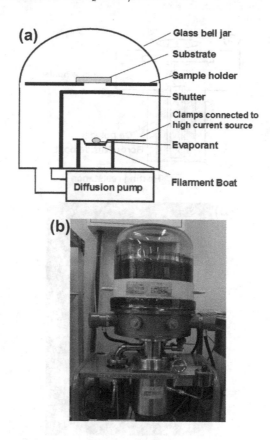

FIGURE 2.10   (a) Schematic and (b) camera photo of metal evaporation system.

In this study, electrodes (mainly Au for the top electrode and Al for the bottom) are formed by metal evaporation. (Author's note: This is the best way for forming high-quality electrodes: easy, fast, and clean.)

2.3 THERMAL DESORPTION CHARACTERIZATION

2.3.1 Thermal Desorption Spectroscopy (TDS)

TDS, also known as temperature programmed desorption (TPD), is one of the most important methods for observing adsorption/desorption of molecules on a solid surface. Figures 2.11(a) and (b) show the schematic and a camera photo of the TDS used in this study. The TDS measurement system with a lamp heating system, a quadrapole mass spectrometer (QMS), a UHV ($10^{-7} \sim 10^{-8}$ Pa) chamber, and a quartz sample holder (EMD-WA1000S/W (ESCO, Ltd.)) were used in a multi-ion detection

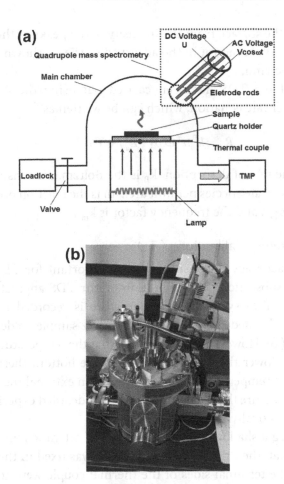

FIGURE 2.11   (a) Schematic and (b) camera photo of TDS system.

(MID) mode with sensitivity up to 10^{-15}A. When molecules come in contact with a surface, they adsorb onto it, minimizing their energy by forming a chemical bond with the surface. The bonding energy varies with the combination of the adsorbate and surface. If the surface is heated at one point, the energy transferred to the adsorbed species will cause it to desorb. The temperature at which this happens is known as the desorption temperature. Since TDS observes the mass of desorbed molecules, it shows what molecules are adsorbed on the surface. Moreover, TDS recognizes the different adsorption conditions of the same molecule from the differences between the desorption temperatures of molecules desorbing different sites at the surface. TDS also obtains the amounts of adsorbed molecules on the surface from the intensity of the peaks of the TDS spectrum, and the total amount of the adsorbed species is shown by the integral of the spectrum.

Analysis of the TDS has been carried out using the Arrhenius or Polanyi-Wigner rate equation, which can be written as[7]

$$R_d = k_m^0 n^m exp\left(-E_a/k_B T\right) \tag{2-1}$$

where R_d is the rate of desorption, k_B is the Boltzmann constant, n is the number of adsorbate species per unit area, m is the reaction order, E_a is the activation energy, and the frequency factor is k_m^0.

2.3.2 Temperature Calibration of TDS

Temperature accuracy is considered very important for TDS measurement. In this subsection, the calibration of our TDS apparatus is introduced. In our TDS system, the temperature is recorded by a bottom thermal couple located close beneath the quartz sample holder (as shown in Figure 2.11(a)). However, it is inevitable that the temperature of sample will be a little lower than that recorded by the bottom thermal couple. To calibrate the temperature precisely, we use an external thermal couple (PtRh-Pt) to measure the real temperature. The detailed experiment setup is shown schematically in Figure 2.12.

First, we dug a shallow hole in the center of a 1cm × 1cm cut Si (100) wafer. After that, the PtRh-Pt thermal couple was fixed in the hole using cement, and the terminal sides of the thermal couple were connected to the vacuum flange. Then, we fixed the flange to the vacuum chamber and started vacuuming the chamber to high vacuum level. And finally, the temperature values from the bottom thermal couple and the external thermal couple were collected separately. Figure 2.13(a) through (d) show a comparison between the temperature values collected from the external thermal couple (T_{ext}) and the bottom thermal couple (T_{bot}) with heating rates of 10, 20, 30, and 60°C/min. Clearly, for the same heating rate, both temperature values show linear profiles with different slopes at a cross point of around 470°C. We can easily determine the slope from Figure 2.13 through the linear fitting.

On the basis of the result in Figure 2.13, the relationship between T_{ext} and T_{bot} can be extracted. Note that for different heating rates, this relationship does not change. And we have

$$T_{ext} = T_{bot} \times 0.6855 + 150.1 \tag{2-2}$$

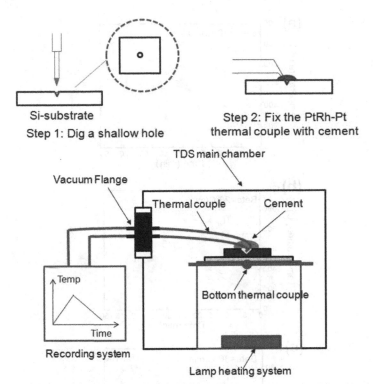

Si-substrate

Step 1: Dig a shallow hole

Step 2: Fix the PtRh-Pt
thermal couple with cement

Step 3: Set the sample into TDS chamber and connect the thermal
couple with vacuum flange

FIGURE 2.12   Schematic of the temperature calibration experiment setup.

Here, we assume that the real temperature of the sample surface is very close to the temperature recorded by the external thermal couple. In this study, for all the TDS measurements, the temperatures were calibrated by using equation (2–2).

2.3.3 Quadrapole Mass Spectrometer (QMS)

Mass spectrometric measuring methods have become indispensable diagnostic aids in numerous branches of process engineering, technology and product development, medicine, and basic scientific research. In contrast to total pressure measurements, in mass selective measuring methods, detection is according to the mass/charge ratio of the ions. Chiefly, quadrupole mass filters (see Figure 2.14) are used today for these measuring tasks. The properties of quadrupole mass filters that are especially useful for these applications are the simple manner of

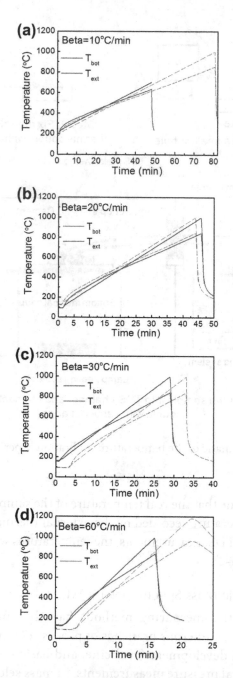

FIGURE 2.13   Temperature calibration of the TDS with sweeping rate of (a) 10, (b) 20, (c) 30, and (d) 60°C/min, respectively. To check the reproducibility, some measurements were carried out again (as seen in the dashed line).

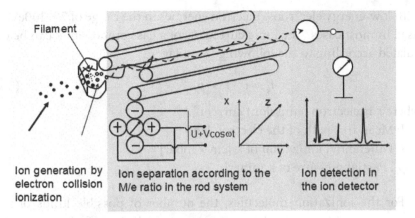

FIGURE 2.14 Schematic of a QMS system.

scanning the entire mass range; high sensitivity; high measuring and repetition rate; large measuring range (up to ten decades); and compatibility with the general requirements of vacuum technology, such as relatively small dimensions, arbitrary mounting position, and low outgassing rates.

The ions are separated in a high-frequency electric quadrupole field between the four rod electrodes with field radius r_0. The voltage between the electrodes consists of a high-frequency alternating voltage $V\cos\omega t$ and a superimposed direct voltage U. When ions are trapped in the direction of the field axis perpendicular to the plane of the diagram, they perform oscillations perpendicular to the field axis under the influence of the high-frequency field. For certain values of U, V, ω, and r_0, only ions with a particular M/e ratio can pass through the separating field and reach the ion detector. Ions that have a different mass/charge ratio are injected by the quadrupole field and therefore cannot reach the detector. Mass/charge ratio scanning can be achieved by varying the frequency ($M/e{\sim}1/\omega^2$) or, as is for technical reasons almost always the case, by varying the voltage ($M/e{\sim}V$). This gives a linear mass scale by simple means. It is also possible to adjust the resolution capability of a quadrupole mass spectrometer via the magnitude U of the direct-voltage component to the amplitude V of the high-frequency component.

Ionization is the part of the procedure for analyzing the neutral particles that has greatest effect upon the sample gas. A fraction of the gas molecules presented in the gas phase are converted into ions by bombarding them with low-energy electrons. Electron energies in the range of 70~100eV are used in most cases. The ionic current i^+_k of a gas component k can be calculated according to the following formula:

$$i^+_k = i^- \cdot I \cdot s \cdot p_k \qquad (2\text{--}3)$$

where i^- is electron (emission) current [A]

I: Mean free path of the electron [cm]

s: Differential ionization of k [cm × mbar]$^{-1}$

p_k: Partial pressure of k [mbar]

For the ionization molecules, the number of possible kinds of ions increases rapidly with the increasing complexity of the molecules. Fragment ions appear in singly and multiply charged molecular ions. The case of GeO(g) desorption is a case in point; the following two reactions usually happen at the same time.

$$GeO + e^- = GeO^+ + 2e^- \qquad (2\text{--}4)$$

$$GeO + e^- = GeO^{2+} + 3e^- \qquad (2\text{--}5)$$

Since Ge has 5 isotopes, ^{70}Ge, ^{72}Ge, ^{73}Ge, ^{74}Ge, and ^{76}Ge, the corresponding mass number of Ge^{16}O should be M = 86, 88, 89, 90, and 92.

2.3.4 Desorption Reaction

Desorption reaction is divided into physical desorption and chemical desorption by their bonding states with the substrate. Consider a molecule [A] adsorption on a substrate. This process contains two stages: 1) physical adsorption and 2) chemical adsorption. The physical adsorption is generally bonded by the Van der Waals force. Desorption of a physically adsorbed molecule does not require the breaking of chemical bonds; it belongs to the physical reaction, so this kind of desorption is regarded as physical desorption. If the molecule acquires enough energy to create a chemical bond, then this molecule enters the next stage as a chemically adsorbate molecule. In this stage, desorption requires the breaking of chemical bonds. Figure 2.15 shows a schematic of physical and chemical adsorption and desorption processes.

For physical desorption, the physical interaction between the adsorbed molecule and substrate is quite small and closely related to the distance from the substrate; therefore, desorption usually occurs at a low

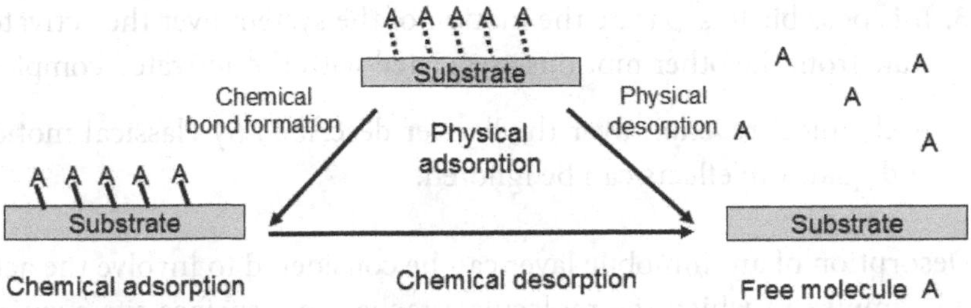

FIGURE 2.15 Schematic of the relationship among chemical adsorption/desorption and physical adsorption/desorption.

temperature. For chemical desorption, the desorption order is derived from the chemical reaction that forms the desorption molecule. For example, consider the following reaction:

$$2A \rightarrow A_2 \qquad (2\text{–}6)$$

The reaction constant k can thus be calculated by considering the equilibrium concentration of the reactants and products:

$$k = \frac{[A]^2}{[A_2]} \qquad (2\text{–}7)$$

Since the product concentration is proportional to the square of the reactant concentration, the order of this desorption is 2.

2.3.5 Desorption Rate

According to conventional transition-state theory, we describe the absolute rates of adsorption and desorption by considering the following assumptions made at equilibrium:[8]

1. Statistically, only a single direction reaction is considered. Reactants that surmount the activation barrier and are heading towards products cannot turn around and form reactants.

2. The reactants have an energy distribution that is described by Maxwell-Boltzmann statistics. It is assumed that even when the whole system is not in equilibrium, the activated complex can be calculated using equilibrium theory.

3. It is possible to separate the motion of the system over the activated state from the other motions associated with the activated complex.

4. A chemical reaction over the barrier described by classical motion and quantum effects can be ignored.

Desorption of an immobile layer can be considered to involve the activated complex in which the molecule attached to a surface site acquires enough energy to escape the surface. Thus, it follows that if the adsorbed and activated complexes are in equilibrium and the adsorbed concentration, n, and the concentration of the activated complex, n*, are in equilibrium, the equilibrium constant K* is given in terms of partition functions. Thus:

$$K^* = \frac{n^*}{n^m} = \frac{q^*}{q} exp(-\frac{E_a}{k_B T}) \qquad (2\text{–}8)$$

where E$_a$ is the activation energy for desorption of a mole of molecules at T = 0K.

The absolute desorption rate is thus given by

$$R_d = n^* \cdot v = n^m \frac{q^*}{q} \frac{k_B T}{h} exp(-\frac{E_a}{k_B T}) \qquad (2\text{–}9)$$

where n is the concentration of the products, q and q* are partition functions related to the reaction, k$_B$ is the Boltzmann constant, h is the Plank constant, E$_a$ is activation energy for desorption, and m is the desorption order.

2.3.6 TDS Analysis

Generally, TDS measurement has two types: the non-isothermal TDS and the isothermal one. For non-isothermal TDS, the temperature is swept by a constant rate, while isothermal TDS has a constant sweeping temperature.

1. Non-isothermal TDS.

For most TDS measurements, the sweeping temperature T is increased by a constant rate β; thus:

$$T = T_0 + \beta t \qquad (2\text{–}10)$$

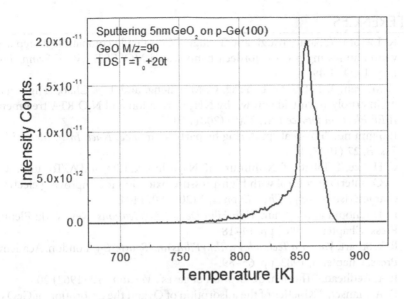

FIGURE 2.16 Typical non-isothermal TDS desorption spectrum. Sample sputtering 5nn GeO$_2$ on p-Ge (100) substrate. Detection mass M/z=90, sweeping rate 20K/min.

where T is the target sweeping temperature, T$_0$ is the initial temperature, β is the sweeping rate, and t is time. In this study, T=0 was the room temperature (300K), and the sweeping rates were 10K/min, 20K/min, 30K/min, 60K/min, and 120K/min. Using non-isothermal TDS, it is possible to identify the desorption mass as a function of temperature. Figure 2.16 shows a typical non-isothermal TDS spectrum with a sweeping rate of 20K/min. This desorption spectrum clearly shows the relationship between the M/z=90 molecules' desorption as a function of temperature.

On the other hand, in most actual thermal processes (e.g., annealing treatment), the temperature is not swept with a constant rate. Linear sweeping is sometimes not enough to provide information about the desorption process as a function of time at the fixed temperature. Therefore, to obtain this kind of information for desorption, isothermal TDS measurement is also an important tool. For isothermal TDS measurement, the temperature is kept constant by varying the annealing time. From the isothermal TDS spectrum, it is possible for us to obtain the information on desorption as a function of time.

REFERENCES

1. K. I. Goto, Y. Sambonsugi, and T. Sugii, "Co salicide compatible 2-step activation annealing process fordeca-nano scaled MOSFETs", *VLSI Symp. Tech. Dig.*, 1999, p. 49.

2. T. M. Pan, T. F. Lei, W. L. Yang, C. M. Cheng, and T. S. Chao, "High quality interpoly-oxynitride grown by NH_3 nitridation and N_2O RTA treatment", *IEEE Electron Device Lett*, **22**(2) (2001) 68.

3. P. Sigmund, "Physical sputtering by particle impact", *Nucl. Instr. Meth. Phys. Res. B*, **27** (1987) 1.

4. C. H. Lee, T. Tabata, T. Nishimura, K. Nagashio, K. Kita, and A. Toriumi, "Ge/GeO_2 interface control with high-pressure oxidation for improving electrical characteristics", *Appl. Phys. Express*, **2** (2009) 071404.

5. K. L. Chopra and I. Kaur, *Thin Film Device Applications*, New York: Plenum Press, Chapter 1, 1983, pp. 14–18.

6. R. V. Stuart, *Vacuum Technology, Thin Films and Sputtering*, London: Academic Press, Chapter 3, 1983, pp. 65–89.

7. P. A. Redhead, "Thermal desorption of gases", *Vacuum*, **12** (1962) 203.

8. D. A. Hansen, "Kinetics of the adsorption of O_2 and the desorption of GeO on Ge (100)", Ph.D. Dissertation, Rensselaer Polytechnic Institute, 1990, p. 33.

CHAPTER 3

Desorption Kinetics of GeO from GeO$_2$/Ge

3.1 INTRODUCTION

The early semiconductor surfaces study mainly focused on the issue of how an oxide layer affected surface conductivity, contact potential, etc.[1] Some of the pioneering experiments determined the thermal restoration temperature,[2] the heat of adsorption of oxygen on germanium,[3] the surface structure of oxygen-covered germanium,[4, 5] and the sticking coefficients of oxygen on germanium surfaces.[6–11] Later studies concentrated on the O$_2$/Ge system as a model for gas-surface interactions.[12–15] In the following decades, the binding geometry of chemisorbed oxygen on germanium was addressed using x-ray photoelectron spectroscopy (XPS) and high-resolution electron energy loss spectroscopy (EELS).[16–21] In the 1990s, the kinetics of the adsorption of O$_2$ and the desorption of GeO on Ge (100) was refreshed by using TDS, beam reflectivity techniques, XPS, and ultra-violet photoelectron spectroscopy (UPS).[22–24]

These earlier studies concentrated mainly on the desorption kinetics of GeO on Ge with a coverage below 1 monolayer (ML). Rosenberg et al.[2] found that upon heating germanium saturated with oxygen, the evolved product, GeO, accounted for the entire uptake of oxygen. Surnev et al.[21] found zero-order kinetics for thermal desorption of GeO from Ge (100) at high oxygen coverage. They found first-order kinetics at low coverage (~0.05 ML) and half-order kinetics at intermediate coverage of approximately 0.12 ML. On the basis of Redhead's line shape analysis theory,[25]

DOI: 10.1201/9781003284802-3

the activation energy values of 51, 52, and 54 ± 2 kcal/mol for the zero, half, and first-order kinetic regions, respectively, were obtained. D. A. Hansen et al.[22, 23] found that for oxygen coverage below 0.07 ML, the desorption kinetics that produce GeO(g) are first order in the coverage of oxygen with an activation energy of E_a= 70kcal/mol. For coverage above 0.07 ML, zero-order kinetics describe the desorption mechanism. The desorption is limited by the rate of desorption of GeO.

Madix et al.[14] used molecular beam relaxation spectroscopy to study the desorption kinetics of GeO on Ge (111) in the temperature range of 750 to 1100K. Electron micrographs of the Ge (111) surface showed triangular indented pits,[26] reflecting its symmetry. A series mechanism was used to model this system, with rapid surface diffusion of the reactants to the edges of pits, where desorption of the reaction product, GeO, occurred. The activation energy for desorption and the pre-exponential parameter were found to be 55 kcal/mol and 7×10^{15} sec^{-1}, respectively. The rate-limiting process was desorption of Volatile GeO. Similar experiments have also been carried out by D. A. Hansen[27] using scanning electron microscopy (SEM); they found four-sided pyramidal pits, which were characterized by (111) facets, from the micrographs of Ge (100) sample after reacting with oxygen. This indicated that desorption of GeO at sites at the edges of the pits and not layer-by-layer removal of the surface layers. Prabhakaran et al.[24] used XPS and UPS to study the GeO desorption from oxygen-enriched Ge (100) and Ge (111). Through *in situ* heating of the oxygen-enriched Ge substrate, a clear core-level peak shift of Ge oxidation state was clearly demonstrated. They found that GeO desorption occurs when the temperature is over 420°C in low-oxygen coverage.

Although the interaction between O$_2$ and Ge has been studied for over four decades, the research object is still limited to 1 ML. Towards the purpose of using GeO$_2$ as a gate dielectric, the GeO$_2$/Ge system should be studied at a wider range. Recently, Kita et al. studied the GeO desorption behavior of Ge (100) substrate covered with a 20nm-thick GeO$_2$ sputtered film.[28] They observed GeO volatilization from GeO$_2$ (25nm)/Ge (20nm, amorphous)/ SiO$_2$ (30nm)/Si when the temperature was higher than 600°C by TDS measurement. But the result of a control group with a structure of GeO$_2$ (25nm)/ SiO$_2$/Si showed no observation of GeO desorption, even at 700°C. They concluded that GeO volatilization is closely related to the GeO$_2$/Ge interface.

Although GeO desorption has been studied widely, the details of the GeO desorption mechanism are still unclear, especially in the case of high oxygen coverage or thick GeO$_2$ on Ge. To control the GeO desorption and for further interface improvement of Ge-based devices, we have to understand the GeO desorption mechanism first. (Author's note: Similar to GeO desorption from the GeO$_2$/Ge system, desorption of intermediate state oxide also exists in many oxides systems: for example, SiO$_2$/SiC, Ga$_2$O$_3$/Ga, Ta$_2$O$_5$/Ta, etc. So the following studies present a common example for revealing the desorption kinetics of these sub-oxides.)

3.2 REVISITING GEO DESORPTION USING THE GE (100) SUBSTRATE

In the TDS experiment carried out by Kita et al.,[28] to avoid the possible desorption signal from the edges and backside of Ge-substrate, they used EB-deposited amorphous Ge film to study the GeO desorption process. That experiment concluded that GeO desorption is closely related to the GeO$_2$/Ge interface, but due to the crystallinity difference between the amorphous Ge and crystalline Ge substrate, the desorption spectrum may be different. Therefore, to further confirm whether there is any difference between GeO desorption from amorphous substrate and crystalline Ge substrate, this subsection discusses a similar experiment that was performed on a crystalline Ge (100) substrate: 22nm-thick GeO$_2$ was deposited on clean Ge (100) and SiO$_2$/Si substrate by sputtering. The physical thicknesses of all the GeO$_2$ layers were confirmed by using grazing incidence x-ray reflectometry (GIXR).

After that, for the GeO$_2$/Ge sample, the native oxide on the back side was removed with a DIW-rinsed dust-free cotton swab, the four sample edges were cut, and fresh Ge substrate edges were left after this treatment. Immediately after that, TDS measurements were performed on both samples with the same sweeping rate of 20°C/ min. The mass numbers of 86, 88, 89, 90, and 92 were considered as the signals for GeO. Figure 3.1 shows the TDS spectra of GeO desorption from both samples.

This experiment reproduces Kita's experiment on crystalline Ge substrate well, and it is consistent with the argument that GeO desorption is closely related to the GeO$_2$/Ge interface.

3.3 GE SUBSTRATE CONSUMPTION DURING DESORPTION

Although it has been asserted that GeO desorption is derived from a GeO$_2$/Ge system while GeO$_2$ remains stable on SiO$_2$ even after receiving high-temperature treatment up to 750°C, there is no direct evidence of the

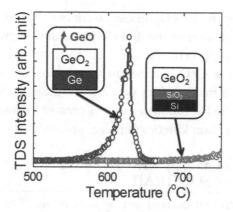

FIGURE 3.1 Intensity of signals corresponding to Ge monoxide (GeO) detected by TDS of the GeO_2/Ge stack and GeO_2/SiO_2 stack. The mass numbers of 86, 88, 89, 90, and 92 were considered as the signals for GeO. Above 520°C, the GeO signal was observed only from the film on Ge, which clearly indicates that the GeO volatilization is driven by the GeO_2/Ge interface reaction. The GeO_2 itself is found to be stable even when the temperature reaches 750°C.

consumption of Ge substrate during GeO desorption process. Therefore, presenting direct evidence of the consumption of Ge substrate will be important to improve our understanding of the GeO desorption process. Moreover, the ratio between Ge substrate and GeO_2 consumed in the GeO desorption is also unclear. The ratio between the reactants is helpful to fix the coefficient in the final chemical reaction that generates GeO(g).

To investigate Ge consumption on a Ge substrate, sputtered GeO_2 fins of 7.5um × 5mm × 110nm (width (W) × length (L) × thickness (t_{ox})) were fabricated using a photolithography technique. The distance between two GeO_2 fins was 1.5um. Figure 3.2(a) shows the schematics of the line patterned GeO_2(110nm)/Ge(100) structure before and after UHV annealing. The as-prepared sample was annealed in UHV (<10^{-7}Pa) from 600 to 750°C until the GeO totally desorbed. After that, the surface and cross-sectional profiles were measured by atomic force microscopy (AFM) with dynamic force mode, as depicted in Figure 3.2(b) and (c), respectively.

After the GeO desorption, the Ge wafer surface just underneath the GeO_2 film was clearly observed to be significantly etched, while the Ge substrate without GeO_2 capping remained a smooth surface. Therefore, we can conclude that the Ge substrate is consumed by the reaction of the GeO

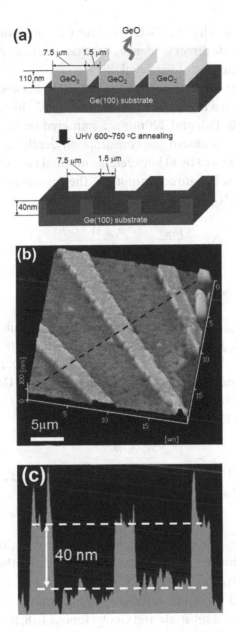

FIGURE 3.2 (a) Schematics of initial and final sample structures, (b) AFM surface image of as-annealed sample, (c) Cross-sectional profile of as-annealed sample.

Source: Reproduced from S. K. Wang et al., "Desorption kinetics of GeO from GeO$_2$/Ge structure", *J. Appl. Phys.* 108, 054104 (2010), with the permission of AIP Publishing.

generation. This is why the GeO₂/Ge interface deteriorates so severely. These results provide direct evidence of the Ge substrate consumption and confirm the conclusions of Kita et al.[28]

To further quantitively probe the relationship between the consumption of Ge atoms from the substrate and the GeO_2 film, GeO_2 fins with thicknesses of 110, 180, and 220nm are sputtered on Ge (100) separately. For example, we can obtain the consumption depth through measuring the step height (denoted by h) between the original Ge surface and the bottom level after Ge substrate consumption. The amount of GeO_2 films N_{GeO2} and Ge substrate (N_{Ge}) can thus be calculated by the following relations:

$$N_{Ge} = \frac{\rho_{Ge} V_{Ge}}{M_{Ge}} = \frac{\rho_{Ge} \cdot W \cdot L \cdot h}{M_{Ge}} \tag{3-1}$$

$$N_{GeO_2} = \frac{\rho_{GeO_2} V_{GeO_2}}{M_{GeO_2}} = \frac{\rho_{GeO_2} \cdot W \cdot L \cdot t_{ox}}{M_{GeO_2}} \tag{3-2}$$

where ρ_{Ge} and ρ_{GeO2} are the physical density of Ge substrate and GeO_2 films; M_{Ge} and M_{GeO2} are the mole mass of Ge substrate and GeO_2 films; V_{Ge} and V_{GeO2} are the volume of Ge substrate and GeO_2 films involved in the reaction; W and L are the width and length of the GeO_2 fins; h is the consumption depth; and t_{ox} is the oxide thickness.

On the basis of (3–1) and (3–2), we have

$$r = \frac{N_{Ge}}{N_{GeO_2}} = \frac{\dfrac{\rho_{Ge} \cdot W \cdot L \cdot h}{M_{Ge}}}{\dfrac{\rho_{GeO_2} \cdot W \cdot L \cdot t_{ox}}{M_{GeO_2}}} = \frac{\rho_{Ge} \cdot h \cdot M_{GeO_2}}{\rho_{GeO_2} \cdot t_{ox} \cdot M_{Ge}} \tag{3-3}$$

Figure 3.3 shows the relationship between consumption depth (h) and the initial GeO_2 thickness (t_{ox}). By taking into account the physical densities of GeO_2 (3.604g/cm³) and Ge (5.323g/cm³),[29] the consumption ratio between GeO_2 and Ge can be calculated by equation (3–3). The consumption ratio between Ge substrate and GeO_2 is about 1.18; this is quite close to 1:1, suggesting that the GeO desorption mainly follows the redox reaction:

$$GeO_2 + Ge \rightarrow 2GeO \tag{3-4}$$

It is noteworthy to point out that GeO desorption might also follows some other reactions, but the main reaction should be the redox reaction

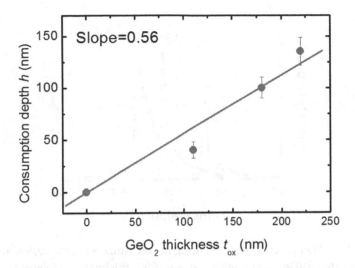

FIGURE 3.3 Relationship between consumption depth and the initial GeO₂ thickness. A slope of 0.56 is obtained by linear fitting.

addressed here. The deviation of the slope from 1:1 might result from the overestimation of the consumption depth during AFM measurement because the depth is quite large.

3.4 DIFFUSION-LIMITED GEO DESORPTION

Figure 3.4 shows the TDS spectra of GeO desorption from GeO₂/Ge with initial GeO₂ thicknesses of 9, 15, and 31nm, respectively (normalized by the surface area). As seen in Figure 3.4, there is a shift from the temperature at which the desorption rate is a maximum, T_p, to higher temperatures with increasing thickness and the leading edge, followed by a sharp drop in the desorption rate.

According to P. A. Redhead's desorption theory,[25] the desorption order should be less than one or the desorption energy increases along with increasing coverage. Some pioneering works by D. A. Hansen and B. Hudson have revealed the desorption kinetics of GeO on Ge (100) for coverage below a mono-layer.[22, 23] They found that at low coverage, desorption followed first-order kinetics with an activation energy of 60 kcal/mol, but at coverage greater than 0.07 ML, zero-order kinetics were observed with the same activation energy. Their results are well explained in terms

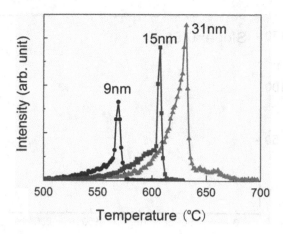

FIGURE 3.4   TDS spectra corresponding to GeO (m/z = 86, 88, 89, 90, 92) from GeO₂ (sputtered)/Ge stacks with various film thicknesses. Sweeping rate is 20°C/min.

Source: Reproduced from S. K. Wang et al., "Desorption kinetics of GeO from GeO₂/Ge structure", *J. Appl. Phys.* 108, 054104 (2010), with the permission of AIP Publishing.

of the instantaneous coverage on active sites for desorption. Therefore, for GeO₂/Ge thicker than several nanometers, the simple Arrhenius or Polanyi-Wigner rate equation is used for spectra fitting. This is written as

$$R_d = k_m^0 n^m exp\left(-E_a/k_B T\right) \qquad (3-5)$$

where R_d is the rate of desorption, k_B is the Boltzmann constant, n is the number of adsorbate species per unit area, m is the reaction order, E_a is the activation energy, and the frequency factor is k_m^0.

For zero-order desorption, m = 0, the desorption rate only depends on the temperature; in other words, at a given temperature, the desorption rate should be the same. However, in Figure 3.4, at the same temperature, the desorption rate decreases as the initial GeO₂ thickness is increased, suggesting the desorption rate does not simply follow the zero-order kinetics.

Figure 3.5 shows the relationship between TDS spectra peak temperature T_p and initial GeO₂ thicknesses ranging from 2 to 33nm for room-temperature ozone-grown, thermally grown (550°C, 1atm), and sputtered GeO₂, respectively.

For the three types of GeO₂, the strong dependence of the peak temperature on the initial oxide thickness is clearly displayed in Figure 3.5.

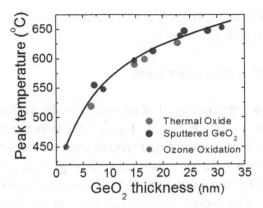

FIGURE 3.5   TDS spectra peak temperature T_p vs. initial GeO_2 thickness ranging from 2 to 33nm for ozone-oxidation, thermal-grown, and sputtered GeO_2, respectively.

Source: Reproduced from S. K. Wang et al., "Desorption kinetics of GeO from GeO₂/Ge structure", *J. Appl. Phys.* 108, 054104 (2010), with the permission of AIP Publishing.

The similar T_p-t_{ox} relationship indicates that GeO desorption is insensitive to the GeO_2 formation methods. The thermal desorption data can be evaluated to extract the thickness dependence of the apparent kinetic parameters for the rate-determining step to form GeO(g). From this relationship, the diffusion process through the oxide film should be considered as the factor governing the desorption temperature. This is consistent with the N_2 ambient annealing results in which the residual GeO_2 thickness shows a non-linear decrease as annealing time increases, from which a diffusion process is considered.[29] Similar results have also been obtained by Y.-K. Sun et al. and Y. Kobayashi et al. from a thin oxide layer on Si (100) and Si (111), respectively.[30-32] Also, Y. Kobayashi et al. looked into the diffusion of SiO formed at the oxide/Si interface through the oxide film.[32-34] Furthermore, if we extend the curve to ultra-thin (<1nm) GeO_2, the desorption peak temperature is estimated to be around 400°C, agreeing well with the results of K. Prabhakaran and T. Ogino.[24]

3.5   ¹⁸O AND ⁷³GE ISOTOPE TRACING IN GEO DESORPTION

Although the thickness dependence of the desorption peaks of GeO is shown in Figures 3.4 and 3.5, it is impossible to conclude from this that the GeO is the diffusion species in this observation. To further study the desorption mechanism, tracing experiments isotopically labeled O$_2$ and Ge in TDS studies will be helpful.

1. ^{18}O isotope tracing experiment.

Before ^{18}O isotope tracing, it is necessary to determine the purity of ^{18}O in the as-prepared GeO$_2$ layer. Figure 3.6 shows the TDS spectra of GeO desorption for the sample of thermally grown 15-nm-GeO$_2$ in 20atm oxygen-18 ambient at 525°C. The similar desorption traces indicate that Ge^{18}O and Ge^{16}O desorb simultaneously. The ratio between ^{18}O tracer and ^{16}O in the GeO$_2$ film grown in oxygen-18 ambient was 0.6:0.4. The incorporation of the oxygen-16 might come from the adsorbed ^{16}O$_2$ or H$_2$^{16}O from the inside wall of the furnace.

Figure 3.7(a) shows the TDS spectra of GeO desorption in Ge^{16}O$_2$(15nm)/Ge^{16}O$_{0.8}$^{18}O$_{1.2}$(15nm)/Ge bilayer structure, and Figure 3.7(b), (c), and (d), respectively, show the corresponding secondary ion mass spectra (SIMS) depth profiles of the as-received sample and those annealed in the TDS

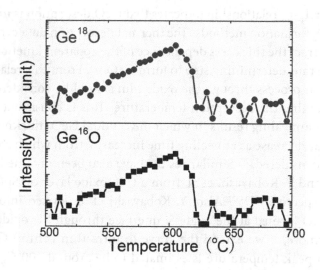

FIGURE 3.6   TDS spectra of GeO desorption for the thermally grown GeO$_2$ in ^{18}O$_2$ ambient. Sweeping rate 60°C/min, mass number 91 for ^{73}Ge^{18}O and 89 for ^{73}Ge^{16}O.

Source: Reproduced from S. K. Wang et al., "Desorption kinetics of GeO from GeO$_2$/Ge structure", *J. Appl. Phys.* 108, 054104 (2010), *Appl. Phys. Lett.* 105, 092101 (2014), with the permission of AIP Publishing.

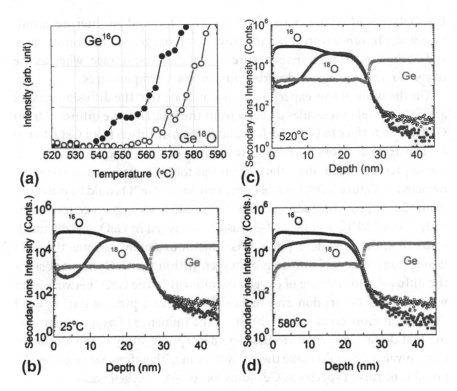

FIGURE 3.7 (a) TDS spectra of GeO desorption in $Ge^{16}O_2/Ge^{16}O_{0.8}{}^{18}O_{1.2}/Ge$ structure, sweeping rate 60°C/min. (b) Depth profiles of the as-deposited sample, (c) those annealed in TDS chamber at 520°C, and (d) 580°C. Note surface profile is subject to SIMS artifact.

Source: Reproduced from S. K. Wang et al., "Desorption kinetics of GeO from GeO₂/Ge structure", *J. Appl. Phys.* 108, 054104 (2010), *Appl. Phys. Lett.* 105, 092101 (2014), with the permission of AIP Publishing.

chamber with terminal temperatures of 520 and 580°C. In Figure 3.7(a), compared with $Ge^{16}O$, the $Ge^{18}O$ desorption shifted to a higher temperature. Only $Ge^{16}O$ desorbs initially, whereas $Ge^{18}O$ is not observed in the lower temperature region below 550°C. Both $Ge^{16}O$ and $Ge^{18}O$ show similar traces in the relatively higher temperature region (>550°C). Moreover, on the basis of the similar depth profiles in Figure 3.7(b) and 3.7(c), we conclude that the bilayer structure is clearly kept, and almost no oxygen intermixing occurs in the bilayer before $Ge^{16}O$ desorbs. These results indicate that the oxygen in the initial $Ge^{16}O$ desorption comes from the top

$Ge^{16}O_2$ layer; otherwise, both $Ge^{16}O$ and $Ge^{18}O$ should be observed simultaneously. In other words, the initially grown oxygen in the bottom layer is desorbed in the high-temperature side of the overall trace, whereas the oxygen in the top layer is desorbed first at lower temperatures.

On the basis of this experiment, we conclude that the diffusion species are not the GeO molecules directly from the interface. Ge diffusion from Ge/GeO_2 interface to GeO_2 surface and oxygen diffusion from GeO_2 to the Ge substrate are possible models to explain our O-18 experiment result. For oxygen diffusion, since the oxygen was totally intermixed at 580°C (as depicted in Figure 3.7(d)), while no desorption of $Ge^{18}O$ could be observed below 550°C, we conclude that the intermixing of oxygen occurs drastically above 550°C, and the self-diffusion of oxygen in GeO_2 layer is much faster than the GeO desorption. It is difficult to simply attribute the GeO desorption to the self-diffusion of oxygen within the GeO_2 network since the diffusive intermixing of oxygen occurs within the GeO_2 network even without GeO desorption and makes the diffusion process during GeO desorption more complex. To eliminate the influence of oxygen intermixing that doesn't contribute to the GeO desorption, study of the motion of oxygen vacancy could make the analysis easier. Therefore, we consider the possible diffusion species as Ge atoms (or ions) or oxygen vacancies.

2. ^{73}Ge isotope tracing experiment.

For the Ge isotope tracing experiment, we chose ^{73}Ge to prepare the isotopically labeled layer. As shown in Figure 3.8, Ge has five isotopes, and the natural abundances of the five Ge isotopes are ^{70}Ge (20.38%), ^{72}Ge (27.31%), ^{73}Ge (7.77%), ^{74}Ge (36.71%), and ^{76}Ge (7.84%). Using the naturally rarest isotope (^{73}Ge) will increase the difference between the isotopically labeled layer and the layer with natural abundance, so as to make the experiment's result clearer. Moreover, ^{73}Ge is the best one for TDS mass number separation in the Ge-O system. Since ^{73}Ge is the only odd-numbered isotope, using the odd-even characteristics to deal a $^{73}Ge+^{16}O+^{18}O$ system is much easier than using other even-numbered Ge isotopes combined with ^{16}O and ^{18}O. However, since ^{73}Ge has the lowest natural abundance, a high purity of ^{73}Ge is necessary for isotope tracing.

To confirm the purity of the ^{73}Ge isotope, a 20-nm thick nominal ^{73}Ge layer was deposited on SiO_2/Si substrate by thermal evaporation in vacuum at 1300°C. After that, the $Ge/SiO_2/Si$ was thermal oxidized in pure O_2

Natural abundances of Ge isotopes

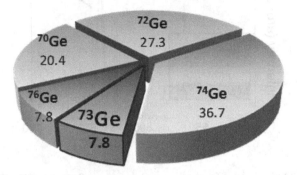

FIGURE 3.8 Natural abundances of the five Ge isotopes are ^{70}Ge (20.4%), ^{72}Ge (27.3%), ^{73}Ge (7.8%), ^{74}Ge (36.7%), and ^{76}Ge (7.8%).

ambient, and finally, we obtained a GeO$_2$/Ge/SiO$_2$/Si stack. As was clarified earlier, GeO desorption occurs by consuming Ge substrate. Therefore, the purity of ^{73}Ge, [^{73}Ge], can be calculated by considering the ratio of ^{73}GeO desorption amount in the total GeO desorption:

$$
\begin{aligned}
\left[^{73}Ge \right] &= \frac{I_{73GeO}}{I_{70GeO} + I_{72GeO} + I_{73GeO} + I_{74GeO} + I_{76GeO}} \\[2mm]
&= \frac{\int_{t_1}^{t_2} R_d * \left(M/z = 89, t \right) \sqrt{T} dt}{\sum_{i=86,88,89,90,92} \int_{t_1}^{t_2} R_d * \left(M/z = i, t \right) \sqrt{T} dt} \\[2mm]
&\approx \frac{\int_{t_1}^{t_2} R_d * \left(M/z = 89, t \right) dt}{\sum_{i=86,88,89,90,92} \int_{t_1}^{t_2} R_d \left(M/z = i, t \right) dt}
\end{aligned}
\tag{3-6}
$$

where I_{70GeO}, I_{72GeO}, I_{73GeO}, I_{74GeO}, and I_{76GeO} in equation (3–6) are the TDS desorption integrated areas ^{70}GeO, ^{72}GeO, ^{73}GeO, ^{74}GeO, and ^{76}GeO, respectively; t_2 is the terminal time for GeO desorption; t_1 is the starting time for GeO desorption; and R_d^* is the detection signal intensity in TDS (I~t) spectrum. The last equation is only possible when t_2 is close to t_1.

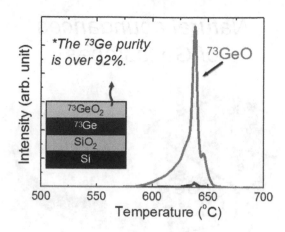

FIGURE 3.9 TDS spectra of GeO desorption from a GeO$_2$/Ge/SiO$_2$/Si stack using the nominal ^{73}Ge powder. The purity of the ^{73}Ge isotope is over 92%.

Figure 3.9 shows the TDS spectra of GeO desorption from a GeO$_2$/Ge/ SiO$_2$/Si stack using the nominal ^{73}Ge powder.

According to equation (3–6), the purity of the Ge powder used in this experiment is about 92%.

Two kinds of samples: ^{Nat}GeO$_2$/^{73}Ge/SiO$_2$/Si, denoted by [A], and ^{73}GeO$_2$/^{Nat}Ge (100) denoted by [B]. For sample [A], a 65nm-thick amorphous ^{73}Ge layer was deposited on SiO$_2$ (1um)/Si substrate by thermal evaporation in vacuum, followed by a 17nm-thick ^{Nat}GeO$_2$ deposition onto the ^{73}Ge layer by sputtering. For sample [B], 12nm-thick ^{73}Ge was deposited by thermal evaporation in vacuum on as-cleaned Ge (100) substrates, followed by annealing in 1atm O$_2$ at 550°C for 15 minutes. The GeO$_2$ thickness was confirmed to be 17nm by the ellipsometry. Comparing the physical densities of GeO$_2$ with amorphous Ge,[35] we concluded that the ^{73}Ge layer was slightly over oxidized, including the ^{Nat}Ge substrate. In addition, the self-diffusivity of Ge in Ge is very small (~2 × 10^{-22} m^2/s) at 550°C;[36] the GeO$_2$ layer in sample [B] is mainly ^{73}GeO$_2$.

Isothermal TDS measurements were performed on samples [A] and [B] at 540°C and 555°C, respectively. For non-isothermal TDS measurements, the temperature was increased from 25°C to 800°C, with a heating rate of 60°C/min. For samples [A] and [B], the mass numbers of M/z = 18, 40, 44, 86, 88, 89, 90, and 92 were chosen, in which M/z = 86, 88, 89, 90, and 92 were considered as the signals from ^{70}Ge^{16}O, ^{72}Ge^{16}O, ^{73}Ge^{16}O, ^{74}Ge^{16}O, and

^{76}Ge^{16}O, respectively. In the ^{73}Ge isotope tracing experiment, the following calculation was applied for the raw data treatment:

$$I_{^{Nat}GeO} = I_{^{70}GeO} + I_{^{72}GeO} + I_{^{73}GeO(Nat)} + I_{^{74}GeO} + I_{^{76}GeO}$$

$$= \frac{I_{^{70}GeO} + I_{^{72}GeO} + I_{^{74}GeO} + I_{^{76}GeO}}{A_{^{70}Ge} + A_{^{72}Ge} + A_{^{74}Ge} + A_{^{76}Ge}} \qquad (3\text{–}7)$$

$$I_{^{73}GeO(pure)} = I_{^{73}GeO(Total)} - I_{^{73}GeO(Nat)} = I_{^{73}GeO(Total)} - I_{^{Nat}GeO}A_{^{73}Ge} \qquad (3\text{–}8)$$

where $I_{^{Nat}GeO}$, $I_{^{70}GeO}$, $I_{^{72}GeO}$, $I_{^{74}GeO}$, and $I_{^{76}GeO}$ in equation (3–7) are the TDS desorption integrated area of GeO, with natural abundance excluding the contribution from the pure ^{73}GeO layer, the GeO desorption integrated areas of ^{70}GeO, ^{72}GeO, ^{74}GeO, ^{76}GeO, respectively. $I_{^{73}GeO(Pure)}$, $I_{^{73}GeO(Total)}$, and $I_{^{73}GeO(Nat)}$ in equation (3–8) are the TDS desorption integrated area of ^{73}GeO from the pure ^{73}GeO$_2$ layer, the total ^{73}GeO desorption integrated area, and the integrated area of ^{73}GeO that comes from the natural abundance–enriched layer. $A_{^{70}Ge}$, $A_{^{72}Ge}$, $A_{^{73}Ge}$, $A_{^{74}Ge}$, and $A_{^{76}Ge}$ are the natural abundance of ^{70}Ge, ^{72}Ge, ^{73}Ge, ^{74}Ge, and ^{76}Ge, respectively. In the oxygen-18 isotope tracing experiment, M/z = 89 and 91, corresponding to ^{73}Ge^{16}O and ^{73}Ge^{18}O, were selected for comparison. During the measurements, the background vacuum level of the main chamber was about $8 \times 10^{-8} \sim 5 \times 10^{-7}$ Pa.

Figures 3.10(a) and (b) show the non-isothermal TDS spectra of GeO desorption for samples [A] and [B], respectively. Using such stacks, we can easily determine where Ge in the desorbed GeO comes from. In Figure 3.10(a), in the initial stage of the measurement (520~530°C), ^{Nat}GeO desorption rate is much higher than that of ^{73}GeO, indicating that Ge in the desorbed GeO predominantly originates from the top ^{Nat}GeO$_2$ layer. A similar result is shown in the reversed structure [B], as depicted in Figure 3.10(b), in which ^{73}GeO shows a higher desorption rate than that of ^{Nat}GeO at the initial stage (550~565°C).

Combined with the O-18 tracing results that oxygen in the desorbed GeO comes from the GeO$_2$ surface, we conclude that in the initial stage of GeO desorption, both Ge and oxygen in the desorbed GeO are initiated from the GeO$_2$ surface. In Figure 3.10(a) and (b), as the sweeping temperature increases, the desorption rate of ^{Nat}GeO becomes closer to that of ^{73}GeO, suggesting that the contribution of Ge from the substrate increases

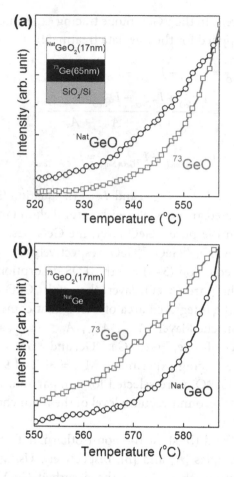

FIGURE 3.10 Non-isothermal TDS spectra of GeO desorption from (a) $^{Nat}GeO_2/^{73}Ge/SiO_2/Si$ (sample [A]) and (b) $^{73}GeO_2/^{Nat}Ge$ (100) (sample [B]).

Source: Reproduced from S. K. Wang et al., "Isotope Tracing Study of GeO Desorption Mechanism from GeO₂/Ge Stack Using ^{73}Ge and ^{18}O", *Japanese Journal of Applied Physics* 50 (2011) 04DA01, with the permission of the Japan Society of Applied Physics.

because it is inferred that the intermixing of Ge between GeO₂ and Ge substrate occurs gradually during the annealing treatment.

The sample structures and the results of the isotope tracing experiments are schematically summarized in Figure 3.11. Figure 3.11 clearly indicates that in the uniform desorption region, both Ge and O in the desorbed GeO come from the GeO₂ surface. Thus, we conclude the diffusion species

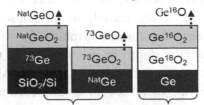

FIGURE 3.11 Schematic sample structures and isotope tracing experimental results.

Source: Reproduced from S. K. Wang et al., "Isotope Tracing Study of GeO Desorption Mechanism from GeO_2/Ge Stack Using ^{73}Ge and ^{18}O", *Japanese Journal of Applied Physics* 50 (2011) 04DA01, with the permission of the Japan Society of Applied Physics.

should not be the directly diffused GeO molecule from the GeO_2/Ge interface. Instead, Ge atoms' (or ions') diffusion from the GeO_2/Ge interface to the GeO_2 surface, or oxygen-vacancies diffusion from the GeO_2/Ge interface to the GeO_2 surface; both are possible explanations for the GeO desorption.

3. Diffusion coefficients comparison between Ge and O.

At the present stage, with only the TDS results using Ge and O isotopes, it is still difficult to clarify what the exact diffusion species is. To further determine the exact diffusion species, a study on the self-diffusion of oxygen and Ge within GeO_2 was carried out.

To determine the diffusion profile of Ge and O upon annealing, $^{Nat}Ge^{16}O_2/^{73}Ge^{18}O_2/SiO_2$/Si, denoted by [C], was used in our experiments. For sample [C], a 20nm-thick amorphous ^{73}Ge layer was deposited on SiO_2 (1um)/Si substrate by thermal evaporation in vacuum, followed by annealing at 525°C in oxygen-18 ambient ($^{18}O_2$ >97%) at 20 atm. The concentration of ^{18}O in the as-grown GeO_2 was estimated to be about 60% by TDS measurement. Here, we define the concentration of ^{18}O as the enrichment of ^{18}O in the GeO_2 layer (^{18}O/ (^{16}O+^{18}O) = 60%). The incorporation of the oxygen-16 might come from the adsorbed $^{16}O_2$ or $H_2^{16}O$ from the inside wall of the furnace. After that, $^{Nat}Ge^{16}O_2$ was deposited by sputtering.

To confirm the thicknesses of the $^{73}Ge^{18}O_2$ and $^{Nat}Ge^{16}O_2$ layers, we deposited them individually onto Si substrates. The thicknesses were confirmed to be about 28nm and 26nm, respectively, by the ellipsometry.

Sample [C] was annealed in 1 atm N_2 ambient for 10 seconds at 600°C and 650°C. After that, the depth profile of Ge and O was measured by the secondary ion mass spectrometry (SIMS). SIMS measurements were performed using a 2.0kV Cs+ ion beam. Five masses—^{28}Si, ^{16}O, ^{18}O, ^{73}Ge, and ^{74}Ge—were simultaneously detected. Since the natural abundance of ^{74}Ge (36.7%) is the greatest among Ge isotopes, and ^{73}Ge is the least, we expected to obtain the Ge diffusion coefficient with a higher accuracy.

Figures 3.12(a) and (b) show the SIMS profile of sample [C] annealed at 600°C and 650°C in 1 atm N_2 for 10 seconds, respectively. In the estimation of the number of diffused isotopes, the profiles near the surface and the GeO₂/SiO₂ interface were not used because the profile in the surface region might suffer from the surface-induced non-equilibrium condition of the measurement, and near the GeO₂/SiO₂ interface region was rather distorted by the matrix effects. We assume that the profile in the region away from 3nm from both the surface and the GeO₂/SiO₂ interface is valid in the following calculations. The ^{18}O and ^{73}Ge fractions at each depth point determined as $^{18}O/(^{18}O+^{16}O)$ and $^{73}Ge/(^{Nat}Ge+^{73}Ge)$ from the results in Figures 3.12(a) and (b) are shown in Figures 3.13(a) and (b), respectively. In Figure 3.13, x = 0 represents the initial interface between $^{Nat}Ge^{16}O_2$ and $^{73}Ge^{18}O_2$. The original point (x = 0) was defined as the point (depth) corresponding to a 50% decrease in the ^{18}O intensity in the SIMS profile of the as-prepared $^{Nat}Ge^{16}O_2/^{73}Ge^{18}O_2/SiO_2/Si$ structure. It is obvious that ^{73}Ge shows a much sharper slope at x = 0 than ^{18}O, strongly indicating that the diffusion coefficient of Ge in GeO₂ is much slower than that of O.

Figure 3.13(a) also shows the depth profile of ^{18}O measured before annealing. The ^{18}O concentration profile of the starting structure in Figure 3.13(a) is a convolution of the real interface with an unknown instrumental broadening function.[37] If the $^{Nat}Ge^{16}O_2/^{73}Ge^{18}O_2$ interface is assumed to be abrupt (a step function) and the broadening function is assumed to be a Gaussian,[38] the resulting convolution will be an error function. Under these assumptions the 2σ width, defined by the positions of 83%–16% of the nominal 60% ^{18}O concentration, is estimated to be about 3nm. If the original $^{Nat}Ge^{16}O_2/^{73}Ge^{18}O_2$ interface does not

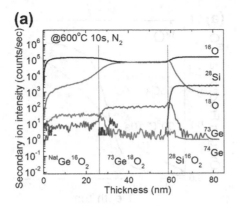

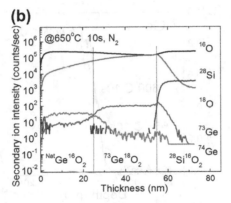

FIGURE 3.12 SIMS profiles of ^{73}Ge and ^{18}O for sample [C] annealed at (a) 600 and (b) 650°C in 1 atm N$_2$ for 10 s, respectively.

Source: Reproduced from S. K. Wang et al., "Isotope Tracing Study of GeO Desorption Mechanism from GeO$_2$/Ge Stack Using ^{73}Ge and ^{18}O", *Japanese Journal of Applied Physics* 50 (2011) 04DA01, with the permission of the Japan Society of Applied Physics.

move and is defined as x = 0, the real ^{18}O and ^{73}Ge concentrations are described by[39]

$$C(x,t) = \frac{1}{2}C_0 erfc\left(\frac{x}{2\sqrt{Dt}}\right) \tag{3-9}$$

where C_0 is the isotope concentration in the starting ^{73}Ge^{18}O$_2$ layer.

By using equation (3–9), we can roughly estimate the diffusion coefficients of Ge and O in GeO$_2$. The calculated results are listed in Table 3.1.

The diffusion coefficient of O is 10~20 times larger than that of Ge at 600 and 650°C. Therefore, the motion of O should be dominant in GeO$_2$.

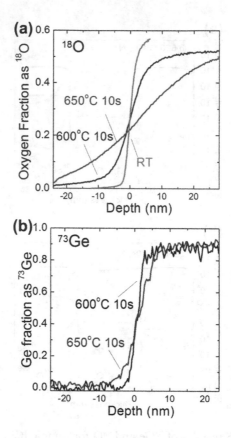

FIGURE 3.13 The concentration-distance relationship of (a) ^{73}Ge and (b) ^{18}O for sample [C] annealed at 600 and 650°C in 1 atm N$_2$ for 10 seconds, respectively; x = 0 represents the initial interface between ^{Nat}Ge^{16}O$_2$ and ^{73}Ge^{18}O$_2$. The original point (x = 0) was defined as the point (depth) corresponding to a 50% decrease in the ^{18}O intensity in the SIMS profile of the as-prepared ^{Nat}Ge^{16}O$_2$/^{73}Ge^{18}O$_2$/SiO$_2$/Si structure.

Source: Reproduced from S. K. Wang et al., "Isotope Tracing Study of GeO Desorption Mechanism from GeO$_2$/Ge Stack Using ^{73}Ge and ^{18}O", *Japanese Journal of Applied Physics* 50 (2011) 04DA01, with the permission of the Japan Society of Applied Physics.

TABLE 3.1 Diffusion Coefficients of Ge and O in GeO$_2$

Temperature (°C)	Ge Diffusion Coefficient D$_{Ge}$ (10^{-18} m^2/s)	O Diffusion Coefficient D$_O$ (10^{-18} m^2/s)
600°C	0.1 ± 0.02	1 ± 0.3
650°C	0.4 ± 0.2	8 ± 2

Furthermore, by taking the surface-initiated desorption into consideration, the diffusion of O must proceed by exchanging with O at its neighbor sites. In this case, the O diffusion process within the GeO$_2$ network can be regarded as the oxygen vacancy diffusion. This result also eliminates the possibility of a paired Ge-O vacancy diffusion because of the large difference between their diffusion coefficients. Thus, we conclude that in the uniform desorption region, the diffusion species for GeO desorption from the GeO$_2$/Ge stack are mainly oxygen vacancies from the GeO$_2$/Ge interface, although Ge diffusion may also provide a little contribution. The diffusion coefficients of Ge and O in GeO$_2$ were estimated on the assumptions that the original $^{Nat}Ge^{16}O_2/^{73}Ge^{18}O_2$ interface (x = 0 before anneal) follows an ideal step function, and the interface does not move during the annealing process. This may result in a little overestimation of the diffusion coefficient, especially for the case of D_{Ge} at 600°C. But this overestimation does not affect the fact that Ge diffuses much slower than O in GeO$_2$.

The oxygen vacancies in this study are assigned to the neutral oxygen mono-vacancy and/or neutral oxygen di-vacancy and/or positively charged vacancy on the basis of ultraviolet absorption spectra analysis and electron-spin resonance (ESR) measurements.[40, 41] It is necessary to point out that further discussion about the oxygen vacancy is not possible for us because distinguishing the neutral oxygen vacancy from the charged one quantitively is quite difficult by experiment. And theoretical study of the oxygen vacancy state in GeO$_2$ is also very rare. However, for the desorption mechanism of GeO from GeO$_2$/Ge, no matter whether the oxygen vacancy is a charged one or a neutral one, we think that our model will not change because the fact that oxygen diffuses much faster than Ge within the GeO$_2$ network does not change.

3.6 REACTIONS AT THE GEO$_2$/GE INTERFACE

To further probe the reaction at the GeO$_2$/Ge interface, a 30nm GeO$_2$ layer was deposited on Ge (100), Ge (111), and Ge (110) substrates by sputtering. Non-isothermal TDS measurements were performed for these three samples at a sweeping rate of 60°C/min. Figure 3.14 shows the GeO desorption results from three kinds of 30nm GeO$_2$/Ge stacks with different substrate orientations. This clearly shows the substrate orientation dependence. For GeO$_2$/Ge (111), the GeO desorption occurs at the lowest rate and the highest peak temperature, while for GeO$_2$/Ge (110), it showed the highest rate

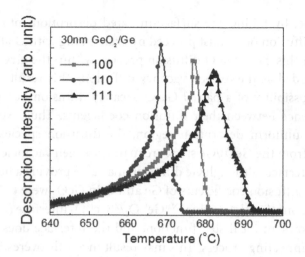

FIGURE 3.14 Non-isothermal TDS spectra of GeO desorption from 30nm GeO₂/Ge stacks with (100), (110), and (111) substrate orientations. The sweeping rate is 60°C/min.

Source: Reproduced from S. K. Wang et al., "Isotope Tracing Study of GeO Desorption Mechanism from GeO₂/Ge Stack Using ^{73}Ge and ^{18}O", *Japanese Journal of Applied Physics* 50 (2011) 04DA01, with the permission of the Japan Society of Applied Physics.

and resulted in a shift in the desorption peak temperature towards lower values.

By comparing the desorption kinetics with the oxidation behavior for three substrate orientations,[42] it is recognized that the GeO desorption from GeO₂/Ge systems is in good agreement with their oxidation behavior. This suggests that the interfacial reaction between GeO₂ and the Ge substrate can be regarded as an oxidation process.

When a Ge substrate is oxidized, such as in back-bond oxygen insertion, the oxygen must come from the GeO₂ network. Thus, the interfacial region at the GeO₂ will always become oxygen-deficient while the partially oxidized Ge can also be treated as an oxygen-deficient network. From the viewpoint of thermodynamic equilibrium, the oxygen-deficient network must include a certain concentration of oxygen vacancies. Thus, as depicted in Figure 3.15, it is inferred that the interfacial redox reaction is the source of oxygen vacancy generation.

Therefore, we regard the interface reaction as redox kinetics. This is consistent with our direct observation of the Ge substrate consumption

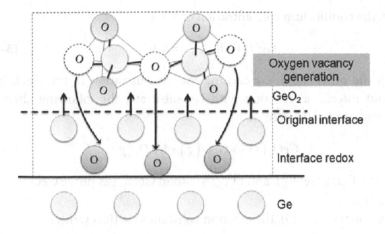

FIGURE 3.15   Schematic of the redox reaction at the GeO$_2$/Ge interface.

Source: Reproduced from S. K. Wang et al., "Isotope Tracing Study of GeO Desorption Mechanism from GeO$_2$/Ge Stack Using ^{73}Ge and ^{18}O", *Japanese Journal of Applied Physics* 50 (2011) 04DA01, with the permission of the Japan Society of Applied Physics.

using line-patterned GeO$_2$/Ge structures with various GeO$_2$ thicknesses. Taking equation (3–4) into consideration, since Ge atoms in the substrate are oxidized by the O atom in the GeO$_2$ network, we infer that O-vacancy (Vo) is generated if there is no additional O incorporation. If we start from this point of view, we can calculate the equilibrium concentration of Vo at the GeO$_2$/Ge interface by thermodynamic calculation. Here we consider two reactions:

$$\frac{1}{2}GeO_2(s)+\frac{1}{2}Ge(s)\rightarrow GeO(g) \qquad (3\text{--}10)$$

$$GeO(g)\rightarrow GeO(V_o) \qquad (3\text{--}11)$$

where GeO$_2$(s) and GeO(g) are solid GeO$_2$ and gas phase GeO. GeO(Vo) is an intermediate state for GeO$_2$ together with an O-vacancy. For equations (3–10) and (3–11), the reaction constants k_1 and k_2 are thus written as:

$$k_1 = \left[p(GeO) \right] \qquad (3\text{--}12)$$

$$k_2 = \left[GeO(V_o) \right] / \left[p(GeO) \right] \qquad (3\text{--}13)$$

Thus, the equilibrium concentration of Vo is

$$[GeO(V_O)]_{GeO_2/Ge} = k_1 k_2 \qquad (3\text{–}14)$$

On the other hand, concerning the Vo generation in the GeO$_2$ bulk without interfacial reaction with Ge substrate, the following chemical reactions are considered:

$$GeO_2(s) \rightarrow GeO(g) + 1/2 O_2(g) \qquad (3\text{–}15)$$

where GeO$_2$(s), GeO(g), and O$_2$(g) are solid GeO$_2$, gas phase GeO, and gas phase O$_2$.

For equation (3–15), the reaction constant k_3 is thus written as:

$$k_3 = \left[p(O_2) \right]^{1/2} \left[p(GeO) \right] \qquad (3\text{–}16)$$

Combining equation (3–14) with (3–16), the equilibrium Vo concentration in GeO$_2$ bulk is:

$$\left[GeO(V_O) \right]_{GeO_2 bulk} = \frac{k_2 k_3}{p(O_2)^{1/2}} \qquad (3\text{–}17)$$

Although we cannot calculate the absolute value of the equilibrium Vo concentration, the relative ratio between the equilibrium Vo concentration at the GeO$_2$/Ge interface and in GeO$_2$ bulk can be obtained from equations (3–14) and (3–17):

$$\frac{\left[GeO(V_O) \right]_{GeO_2/Ge}}{\left[GeO(V_O) \right]_{GeO_2 bulk}} = \frac{p(O_2)^{1/2} k_1}{k_3} \qquad (3\text{–}18)$$

In equation (3–18), the reaction constants k_1 and k_3 are directly obtained from the thermodynamic database.[43] Therefore, the Vo concentration ratio is only a function of the O$_2$ pressure and the temperature. By plotting the Vo concentration ratio in equation (3–18) as a function of the temperature, it is clear that the Vo concentration at the GeO$_2$/Ge interface is much higher than that in GeO$_2$ bulk, as shown in Figure 3.16. Due to a large difference between the equilibrium Vo concentration at the GeO$_2$/Ge interface and in GeO$_2$ bulk, it is expected that there should be a diffusion flux of Vo from the interface to the bulk. This result is consistent with our observation of the sub-gap photo-absorption in annealed GeO$_2$/Ge in N$_2$ at 600°C where Vo was considered to be generated at the interface.[44]

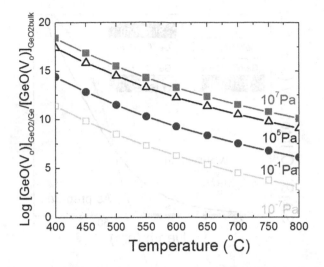

FIGURE 3.16 Vo concentration ratio between GeO_2/Ge interface and GeO_2 as a function of temperature with O_2 pressure at 10^7, 10^5, 10^{-1}, and 10^{-7} Pa. The Vo concentration at GeO_2/Ge interface is much higher than that in the GeO_2 bulk.

Thus, on the basis of thermodynamic calculation, it is concluded that the GeO_2/Ge interface should be regarded as a source of Vo generation.

3.7 EVIDENCE OF VO CONSUMPTION DURING GEO DESORPTION

On the basis of the isotope tracing experiments, we found that the desorbed GeO was derived mainly from the GeO_2 surface. To establish the connection between the interfacial reaction and the surface GeO desorption, we infer that GeO desorption results from the diffusion of Vo, and surface GeO desorption is thus reasonably regarded as a Vo consumer.

To further confirm this model, a bi-layer stack, $Ge^{16}O_2$ (40nm)/$Ge^{18}O_2$ (15nm)/Ge, was prepared by 530°C thermal oxidation in $^{18}O_2$ ambient, followed by sputtering $Ge^{16}O_2$. This bi-layer stack was partly capped with 15nm-thick Si using electron-beam deposition. The schematic is shown in the inset of Figure 3.17. Immediately after that, the as-prepared samples were annealed in N_2 at 550°C for 30 seconds. After that, the O profiles were measured by secondary SIMS. Four masses, ^{28}Si, ^{16}O, ^{18}O, and ^{74}Ge, were simultaneously detected. The SIMS profile of ^{18}O distribution is shown in Figure 3.17.

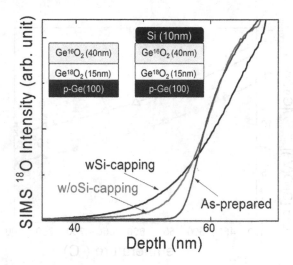

FIGURE 3.17   The SIMS profile of ^{18}O distribution in Si/Ge^{16}O$_2$/Ge^{18}O$_2$/Ge and Ge^{16}O$_2$/Ge^{18}O$_2$/Ge stacks annealed in N$_2$ ambient at 550°C for 30 seconds. The sample without capping shows a sharper oxygen diffusion slope than the one with Si capping.

Source: Reproduced from S. K. Wang et al., "Kinetic Effects of O-Vacancy Generated by GeO₂/Ge Interfacial Reaction", *Japanese Journal of Applied Physics* 50 (2011) 10PE04, with the permission of the Japan Society of Applied Physics.

According to a previous study using Si, HfO₂ as a capping layer to suppress GeO desorption, an Si capping layer is much better at suppressing GeO desorption from the surface.[28] Therefore, it is reasonable to believe that 15nm-thick Si is more effective to reduce the GeO desorption than no capping layer. The sample without capping shows a sharper O diffusion slope than the one with Si capping. This result can be well explained by the Vo diffusion model as follows. Since Vo is consumed at the surface, the residual Vo concentration in the uncapped stack is lower than that in the capped one. Therefore, a lower exchange rate of O in the GeO₂, namely a sharper slope in Figure 3.17, is obtained.

3.8 NONUNIFORM GEO DESORPTION

For SiO₂/Si, the spatial inhomogeneity in the decomposition of ultrathin oxide on Si has been investigated experimentally, and void-formation kinetics were developed.[30–34] To find out more about this on GeO

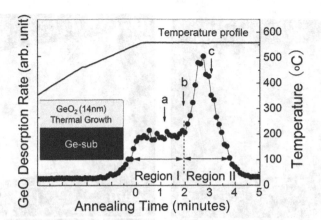

FIGURE 3.18 Isothermal TDS spectra of GeO desorption from GeO$_2$/Ge (100) at 555°C. Note that a, b, and c correspond to the middle of region I, the critical point between region I and II, and the peak in region II.

Source: Reproduced from S. K. Wang et al., "Desorption kinetics of GeO from GeO$_2$/Ge structure", *J. Appl. Phys.* 108, 054104 (2010), with the permission of AIP Publishing.

desorption from GeO$_2$/Ge, the isothermal TDS experiment was done by changing the annealing time in UHV. Figure 3.18 shows the TDS spectra of GeO desorption from GeO$_2$ (14nm, thermally grown)/Ge (100) annealed at 555°C in TDS chamber (<10^{-7}Pa). The desorption spectra can be divided into two regions. One is the initial stage (region I), and the other is the peak with a relatively high desorption rate (region II).

Figures 3.19(a), (b), and (c) show AFM pictures of GeO$_2$/ Ge (100) films annealed at 555°C in UHV for one, two, and three minutes, which correspond, respectively, to the middle of the flat stage, the critical point between the two regions, and the desorption peak in region II. The surface slightly roughens, but no voids are observed in the initial phase, suggesting that the flat stage may correspond to the relatively uniform desorption of GeO. In Figure 3.19(b), several voids are locally observed, whereas the rest of the surface remains smooth, indicating that the desorption peak in Figure 3.18 corresponds to the formation and growth of voids and finally results in a relatively rough surface as shown in Figure 3.19(c).

Concerning Figure 3.18, in region I, the GeO desorption is considered to occur uniformly. For the desorption peaks in region II of the isothermal TDS spectra, however, it is attributed to the nonuniform desorption with

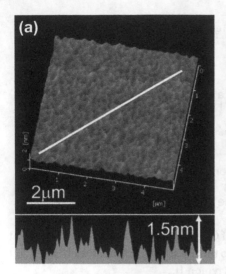

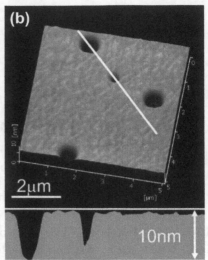

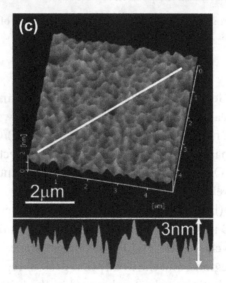

FIGURE 3.19 Surface pictures and cross-sectional profiles of GeO₂(14nm)/Ge (100) with annealing times of (a) 1, (b) 2, and (c) 3 minutes in UHV, corresponding to (a), (b), and (c) from the spectrum in Figure 3.18.

Source: Reproduced from S. K. Wang et al., "Desorption kinetics of GeO from GeO₂/Ge structure", *J. Appl. Phys.* 108, 054104 (2010), with the permission of AIP Publishing.

void formation from the AFM observations. To further discuss this result, we define α as

$$\alpha = I_N/I_0 \tag{3–19}$$

to evaluate the nonuniform degree of GeO desorption, where I_N and I_0 are the integrated areas of region II and the total area, respectively. Figure 3.20(a) shows the isothermal TDS spectra of GeO desorption from sample [A], in which regions I and II are divided by broken lines; a as a

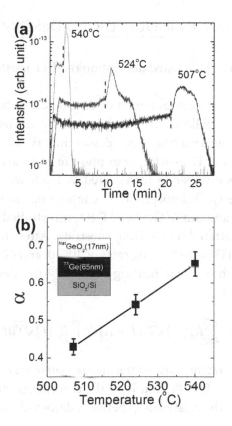

FIGURE 3.20 (a) Isothermal TDS spectra of GeO desorption from ^{Nat}GeO$_2$/^{73}Ge/SiO$_2$/Si stacks at 540, 524, and 507°C. Regions I and II are divided by broken lines. (b) α as a function of annealing temperature.

Source: Reproduced from S. K. Wang et al., "Desorption kinetics of GeO from GeO$_2$/Ge structure", *J. Appl. Phys.* 108, 054104 (2010), *Appl. Phys. Lett.* 105, 092101 (2014), with the permission of AIP Publishing.

function of annealing temperature is shown in Figure 3.20(b). This indicates that the lower temperature is more likely to follow the uniform-desorption mechanism (a smaller α). The temperature-dependent tendency of α suggests that the uniform desorption and nonuniform desorption originate from different mechanisms. Since voids are locally observed in the latter region of GeO desorption, it is easy to connect the formation of these voids to some nucleation process. Concerning the origin of the voids, it seems difficult to explain it by just considering the incorporation of oxygen vacancy into the GeO₂ network; the detailed mechanism will be discussed in the next chapter.

3.9 A VO DIFFUSION-INDUCED GEO DESORPTION MODEL

1. Elimination of H-assisted desorption or H₂O assisted mechanism.

On the basis of the aforementioned discussions, it is believed that GeO desorption mainly follows an oxygen vacancy diffusion mechanism. Some groups argue that the GeO desorption is dominated by other mechanisms, such as H-assisted desorption or H₂O-assisted mechanism because the mass number 18 (considered as H₂O) was always observed together with the GeO desorption. To confirm our model, the relationship between the amount of GeO and GeO₂ was studied. GeO₂ in various thickness was sputtered on Ge (100) substrate. After that, these samples were heated in UHV with heating rates of 20, 30, and 60°C/min. For TDS measurement with a linear heating rate, the total desorption amount I can be written as

$$I = AS_0 \int_{T_0}^{T} R_d(T)\sqrt{T}dT = A\beta S_0 \int_{0}^{t} R_d(t)\sqrt{T}dt \qquad (3\text{--}20)$$

where A is the pre-factor, S_0 is the normalized surface area, β is the heating rate that meets $T = T_0 + \beta t$, and R_d is the desorption rate (molecule/sec). Figure 3.21 shows the relationship between desorbed GeO and the initial GeO₂ thickness.

The linear relationship clearly indicates that the desorption amount of GeO is proportional to the initial GeO₂ amount, suggesting that during the GeO desorption, GeO₂ consumption follows only one dominated reaction. Moreover, the desorption signal integration ratio between H₂O and GeO was also investigated.

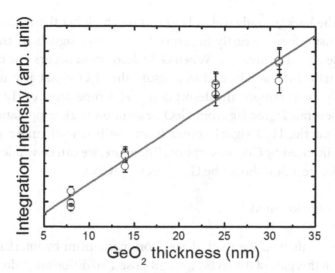

FIGURE 3.21 GeO desorption amount vs. initial GeO₂ thickness ranging from 7nm to 33nm for GeO₂/Ge (100).

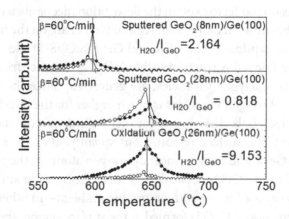

FIGURE 3.22 TDS spectra of H_2O and GeO desorption from 8nm and 28nm sputtered and 26nm thermal grown GeO₂/Ge (100) structure. H_2O: black dot; GeO: white dot.

Figure 3.22 shows the integration area ratio between H_2O and GeO under different conditions. Obviously, the large variation in the ratio indicates that the reliability of the H_2O signal was too poor to be used for analysis. We infer the H_2O intensity was mainly contributed by the pressure effect. In other words, the H_2O was mainly a fake signal.

Since the background level of H_2O was much higher than other signal, H_2O intensity is more easily influenced than other signals by the variation of the chamber pressure. When GeO desorption occurs, the chamber pressure will shift rapidly, and as a result, the H_2O signal also increases drastically. Furthermore, the desorption peak temperature of H_2O always shows a several-degree lag from GeO desorption peak temperature, suggesting that the H_2O signal comes from the by-the-chamber pressure variation induced by GeO desorption. Therefore, we can conclude that the domination reaction should be $Ge + GeO_2 \rightarrow 2GeO$.

2. Vo diffusion model.

Having confirmed the main desorption mechanism by interface redox reaction, in this part of the section, we propose a Vo diffusion–induced GeO desorption model to explain the GeO desorption mechanism from GeO_2/Ge. GeO desorption in region I can be well described by taking oxygen vacancy diffusion into consideration. In this section, we present an oxygen-vacancy diffusion model to explain the desorption mechanism of GeO from GeO_2/Ge in the uniform desorption region. First of all, on the basis of comparing GeO desorption in GeO_2/Ge and GeO_2/SiO_2/Si stacks, we suppose there is a reversible redox reaction at the GeO_2/Ge interface from which the diffusion species, oxygen vacancies, are generated. Since the equilibrium concentration of oxygen vacancies is much higher for the interfacial region than for the GeO_2 bulk, the oxygen vacancies diffuse across the GeO_2 film and reach the GeO_2 surface region. Consequently, after the surface reaction between GeO_2 and oxygen vacancy, oxygen atoms in the reactants are trapped back in the GeO_2 network and form GeO at the surface region. Since GeO_2 is exposed to vacuum or another ambient—in other words, it is an open system—once GeO is formed at the surface region, the desorption happens, and this is not a reversible reaction.

The schematic of this GeO desorption model is illustrated in Figure 3.23. Using this model, we can explain the entire phenomenon observed so far, such as the thickness dependence of desorption temperature, sub-gap formation, and annihilation.[29]

3.10 KINETIC CALCULATION OF GEO DESORPTION FROM GEO₂/GE

In this section, we concentrate on the activation energy of GeO desorption on the basis of the kinetic model shown earlier. Figure 3.18 clearly

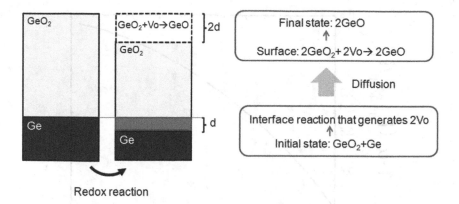

FIGURE 3.23 Schematic of the GeO desorption mechanism from GeO₂/Ge stacks in the uniform desorption region. Redox reaction at GeO₂/Ge interface generates diffusion species, oxygen vacancy (Vo) at the interface, which triggers Vo diffusion in film. Diffusion of Vo in GeO₂ will lead to an oxygen-poor region at the GeO₂ surface where volatile GeO is generated. From the viewpoint of the net consumption amount, once one unit amount of Ge becomes GeO₂, it generates two unit amounts of GeO at the GeO₂ surface.

shows that desorption of GeO follows different mechanisms for different regions. For region II, the GeO desorption rate is much faster and closely related to the lateral growth of the holes. In the present stage, it is difficult to quantitatively analyze the GeO desorption behavior in region II. However, in region I, it is only dominated by the diffusion limited process. Therefore, quantitative calculation might be available in this area since the non-isothermal TDS was performed from a much lower temperature where no desorption can be observed to a certain temperature T_a, we can define the reaction ratio α as

$$\alpha = \frac{I_1}{I_0} = \int_{T_0}^{T_a} R_d dT \Big/ \int_{T_0}^{T_f} R_d dT \qquad (3\text{-}21)$$

where I_1 and I_0 are the integration areas within the sections of $[T_0, T_a]$ and $[T_0, T_f]$, respectively. R_d is the desorption rate of GeO. T_f is the temperature at which no desorption can be observed any more.

The reaction ratio α is the percentage of the molecule desorbed below $T\alpha$. Moreover, at a given rate, we can safely assume that in the initial 10%, the desorption should be dominated by the uniform desorption. GeO₂/Ge isothermal TDS spectra with various thicknesses were analyzed by

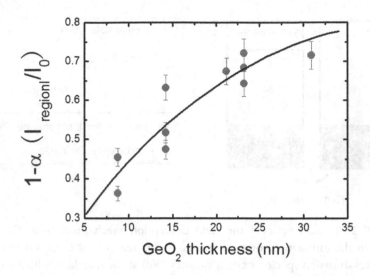

FIGURE 3.24 The relationship between integration area ratio of $I_{\text{region I}}/I_0$ and the initial GeO₂ thickness. Here, $I_{\text{region I}}$ and I_0 represent the integration area of region I and the total integration area of GeO desorption.

comparing the integration of the initial, relatively flat region (region I) and the peak region (region II), and it was found that the ratio of the integration area of region I shows thickness dependence, as depicted in Figure 3.24. In thinner GeO₂ cases (~5nm), region I stands for a relatively small amount but usually larger than 30%.

Thus, below 10%, we consider the simple diffusion process from interface to surface. There are three flux, F_1, F_2, and F_3, which correspond to the interface reaction, diffusion through GeO₂, and the surface reaction, respectively (as depicted in Figure 3.25). We can suppose in a steady state condition

$$F_1 = F_2 = F_3 = R_d \tag{3-22}$$

and

$$F_1 = k_{int}C_{int} \tag{3-23}$$

$$F_2 = D\frac{C_{int} - C_s}{t_{ox}} \tag{3-24}$$

$$F_3 = k_s C_s \tag{3-25}$$

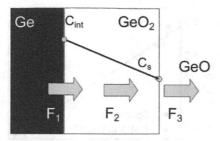

FIGURE 3.25 Schematics of the diffusion process for GeO desorption. F_1, F_2, and F_3 are fluxes that corresponding to interface reaction, diffusion through GeO$_2$, and surface reaction, respectively. C_{int} and C_s are the interface and surface concentrations of diffusion species, respectively.

Source: Reproduced from S. K. Wang et al., "Desorption kinetics of GeO from GeO$_2$/Ge structure", *J. Appl. Phys.* 108, 054104 (2010), with the permission of AIP Publishing.

where k_{int} and k_s are the interface and surface reaction constants, D is the diffusion coefficient of Vo within GeO$_2$, and C_{int} and C_s are the interface and surface Vo concentrations, respectively.

By combining equations (3–22), (3–23), (3–24), and (3–25), we have

$$R_d = D\left(C_{int} - C_s\right)t_{ox}^{-1} \approx DC_{int}t_{ox}^{-1} \ (C_{int} \gg C_S) \qquad (3\text{–}26)$$

Since the desorption is mainly limited by the diffusion process, we assume the diffusion coefficient and the interface concentration can be extended as

$$D = D_0 exp\left(-E_a/k_B T\right) \qquad (3\text{–}27)$$

$$C_{int} = C_0 exp\left(-E_{int}/k_B T\right) \qquad (3\text{–}28)$$

where D_0 is the pre-factor for the diffusion coefficient, E_a is the activation energy for diffusion, k_B is the Boltzmann constant, C_0 is the concentration pre-factor, and E_{int} is the activation energy for interface reaction.

From (3–26), (3–27), and (3–28), we have

$$R_d = D_0 C_0 t_{ox}^{-1} exp(-\frac{E_a + E_{int}}{k_B T}) \qquad (3\text{–}29)$$

By applying equation (3–29) to the present TDS analysis, the desorption activation energy ($E_a + E_{int}$) of about 2eV can be extracted from the slope by plotting ln(t_{ox}) against 1/Tα=0.1, as shown in Figure 3.26.

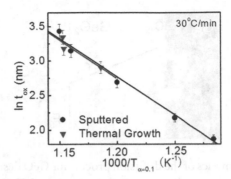

FIGURE 3.26 Relationship between desorption temperature ($T\alpha=0.1$) of GeO and thicknesses for thermally grown GeO_2 and sputtered GeO_2. Sweeping rate is 30°C/min. Activation energy of about 2 eV can be extracted from slope.

Source: Reproduced from S. K. Wang et al., "Desorption kinetics of GeO from GeO₂/Ge structure", *J. Appl. Phys.* 108, 054104 (2010), with the permission of AIP Publishing.

Compared with the Ge-O bonding energy in a Ge-O network, this activation energy for diffusion is quite reasonable when an O atom diffusion in a solid mediated by Vo is considered.

By applying the kinetic model and parameters obtained in this chapter, it is possible to simulate the desorption rate of GeO from GeO_2/Ge as a function of temperature. Figure 3.27 shows a simulation result of GeO desorption TDS spectrum from an 8nm-thick GeO_2 on Ge (100). The simulation result fits well with the experimental data, suggesting that the kinetic model proposed in this study is quite applicable for understanding the GeO desorption behavior in the initial region.

3.11 DISPROPORTIONATE REACTION IN THE GEO₂/GE SYSTEM

(Author's note: In 2012, the author published a paper about GeO disproportionation in *Appl. Phys. Lett.* 101, 061907 (2012). Therefore, most content in this section has been updated.)

Although we consider that Vo is generated from the interfacial redox reaction, from the view point of thermodynamic calculation, the free energy of the interfacial reaction $GeO_2(s) + Ge(s) \rightarrow 2GeO(s)$ as a function of temperature is plotted in Figure 3.28 using HSC Chemistry® 6.0 database (OUTOKUMPU Tech).

In Figure 3.28, it is clear that free energy is always a positive value for the interface reaction. It means the formation of GeO will increase the

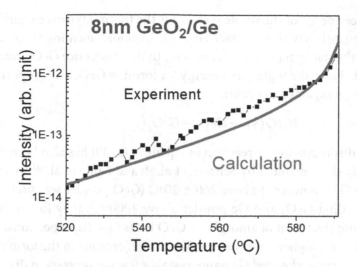

FIGURE 3.27 Simulation result of GeO desorption TDS spectrum from a 8nm-thick GeO₂ on Ge (100) using the Vo diffusion model proposed in this study. The simulation result fits well with the experimental data.

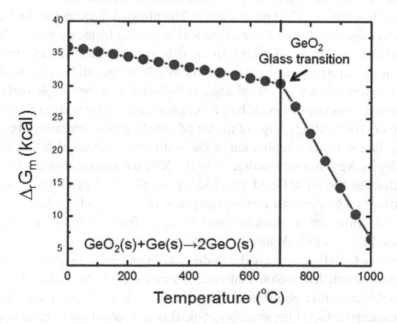

FIGURE 3.28 Gibbs free energy difference between product and precursors ($\Delta_r G_m$) in the interfacial reaction as a function of temperature. The turning point at 700°C is attributed to the glass transition (T_g) temperature of GeO₂.

total free energy of the whole system for the thermodynamic equilibrium case (regardless of the interface energy). Therefore, forming GeO_2 and Ge will be the stable matters in this system. In this case, once GeO is instantly formed, due to the higher free energy for forming GeO, a disproportionate reaction is expected to occur:

$$2GeO(s) \rightarrow Ge(s) + GeO_2(s) \tag{3-30}$$

The disproportionate reaction in equation (3–30) has also been experimentally demonstrated by Schacht et al.[45] and Sahle et al.[46] By annealing the GeO powder at above 260 ± 20°C (GeO powder was prepared by co-annealing GeO_2 and Ge powder above 1000°C), they found that the annealing treatment of amorphous GeO leads to a disproportionate reaction and, at a higher temperature (509 ± 15°C), results in the formation of oxide matrix–embedded Ge nanocrystals a few nanometers in diameter.

To demonstrate the disproportionate reaction in thin film stacks, a mixture of Ge and GeO_2 powders with high purity (>99.9%) was used as the starting material for GeO synthesis, and (1um) SiO_2/Si (100) was used as substrate. First, the mixture was inserted into a boron nitride cell that was prepared for ultra-high vacuum (UHV) thermal evaporation. The cell was then heated to 680°C for GeO generation.[47] The physical thickness of the GeO film was measured to be about 57nm with a grazing incidence x-ray reflectometer. Immediately afterward, the as-deposited sample was subjected to XPS using an Al K_α source (hv = 1486.7 eV). Note that all the data in this section were taken at a take-off angle of 90° relative to the sample surface. Photoelectrons were collected by a hemispherical analyzer with pass energy of 10 eV. The binding energy of the Ge 3d core level was determined by the Voigt line shape (a combination of Gaussian and Lorentzian) fitting after Shirley background subtraction.[48] In the XPS measurement of the Ge 3d spectra, the spectra of Ge $3d_{5/2}$ and $3d_{3/2}$ generally overlapped, with energy splitting of 0.6 eV and an area intensity ratio of Ge $3d_{5/2}$:$3d_{3/2}$ = 3:2.[49] Since the GeO_x films are too thick to detect the signal from the SiO_2/Si substrate, we used C1s = 285.0 eV instead as a reference for calibration. Figure 3.29 shows the Ge 3d spectrum of the as-deposited GeO/SiO_2/Si sample. Clearly, the peak is well deconvoluted into two symmetric peaks located at 31.0 and 31.6 eV, which correspond to Ge^{2+} $3d_{3/2}$ and Ge^{2+} $3d_{5/2}$, indicating that highly stoichiometric GeO film was fabricated. This is consistent with results presented in the literature.[48, 49] According to our TDS study of a 13.4nm-thick GeO/SiO_2/Si sample under UHV (<10⁻⁷Pa), as illustrated in Figure 3.30,

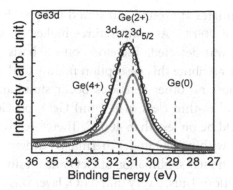

FIGURE 3.29   Ge 3d XPS spectrum of an as-prepared GeO (57nm)/SiO₂/Si stack; the peak at around 31.3 eV is assigned to Ge²⁺. Dashed lines indicate the positions of Ge⁰⁺ and Ge⁴⁺.

Source: Reproduced from S. K. Wang et al., "Kinetic study of GeO disproportionation into a GeO₂/Ge system using x-ray photoelectron spectroscopy", *Appl. Phys. Lett.* 101, 061907 (2012), with the permission of AIP Publishing.

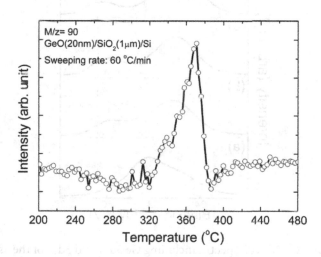

FIGURE 3.30   TDS desorption spectrum of an as-prepared GeO (13.4nm)/SiO₂/ Si stack. Note that M/z=90 is selected for detection, and the sweeping rate is 60°C/min.

Source: Reproduced from S. K. Wang et al., "Kinetic study of GeO disproportionation into a GeO₂/Ge system using x-ray photoelectron spectroscopy", *Appl. Phys. Lett.* 101, 061907 (2012), with the permission of AIP Publishing.

the desorption initiates at around 320°C and reduces drastically after the desorption peak at 400°C. At temperatures higher than 400°C, no GeO desorption signal was detected. Therefore, on the basis of the TDS result in Figure 3.30, we attribute this desorption mainly to the self-sublimation of GeO, and almost no other reactions occur simultaneously. For GeO desorption from ultra-thin (less than 1nm) GeO_2 on Ge, the desorption temperature should be no less than 400°C. Therefore, to further study the possible reactions within GeO under thermal treatment, a capping layer is required to avoid the self-sublimation of GeO and attain a thermodynamic equilibrium condition. Thus, a very thin Al_2O_3 layer was deposited onto the as-prepared GeO surface through magnetron sputtering. The thickness of the Al_2O_3 layer was estimated to be less than 2nm by XPS.

Figures. 3.31(a) through (d) show the Ge 3d core-level spectra (including $3d_{3/2}$ and $3d_{5/2}$) of the as-prepared Al_2O_3/GeO/SiO₂/Si sample and

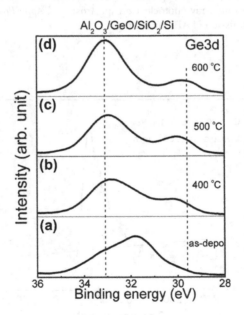

FIGURE 3.31   Ge 3d XPS spectra including Ge $3d_{3/2}$ and $3d_{5/2}$ of the as-prepared Al_2O_3/GeO/SiO₂/Si stack and the stacks annealed at 400°C, 500°C, and 600°C for 30 minutes. Compared to the result in Figure 3.29, the as-prepared stack with Al_2O_3 capping has a shoulder at about 33.0 eV. This suggests that the GeO film is partially oxidized during Al_2O_3 deposition.

Source: Reproduced from S. K. Wang et al., "Kinetic study of GeO disproportionation into a GeO₂/Ge system using x-ray photoelectron spectroscopy", *Appl. Phys. Lett.* 101, 061907 (2012), with the permission of AIP Publishing.

the samples annealed at 400°C, 500°C, and 600°C for 30 minutes under a UHV condition, respectively. For the as-prepared sample, as shown in Figure 3.31(a), there is only one asymmetric peak, around 31.7eV with a shoulder on the higher binding-energy side, suggesting that the as-prepared sample mainly contains GeO and is partially oxidized to a higher oxidation state by the surface sputtering of the Al_2O_3 capping layer. After the UHV annealing treatment, there is a clear shift of the peak corresponding to Ge^{2+} towards the higher and lower binding-energy sides, which provides strong experimental evidence of the disproportionation reaction $2GeO \rightarrow GeO_2 + Ge$ after UHV annealing treatment.

Moreover, this result is confirmed by Raman spectroscopy. As shown in Figure 3.32, for the as-deposited sample, almost no signal related to Ge is observed. As the annealing temperature increases, there is a broad band that corresponds to the first-order optical phonon of amorphous Ge at around 280 cm⁻¹. When the annealing temperature reaches 600°C, besides the broad band, there is a sharp band at around 300 cm⁻¹, indicating that Ge is partially crystallized because the sharp band is attributed to the

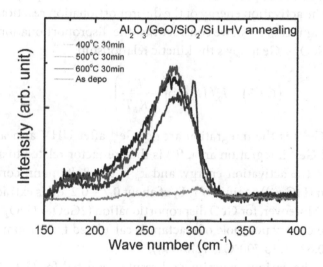

FIGURE 3.32 Raman spectra of the as-prepared Al_2O_3/GeO/SiO_2/Si stack and the stacks annealed at 400°C, 500°C, and 600°C for 30 minutes. The broad bands around 280 and 300cm⁻¹ are assigned to the first-order optical phonons of amorphous and crystalline Ge, respectively.

Source: Reproduced from S. K. Wang et al., "Kinetic study of GeO disproportionation into a GeO₂/Ge system using x-ray photoelectron spectroscopy", *Appl. Phys. Lett.* 101, 061907 (2012), with the permission of AIP Publishing.

first-order optical phonon of crystalline Ge.[50] This observation is consistent with the thermodynamic calculation and other reports. Therefore, we conclude without any ambiguity that there is GeO disproportionation into GeO_2 and Ge after UHV annealing.

Furthermore, for a fixed annealing time span (30 minutes), as the annealing temperature increases, the contribution of Ge^{2+} decreases. This makes it possible to extract the activation energy of GeO disproportionation by plotting the peak area of the Ge^{2+} part against temperature. Therefore, to obtain the exact area of the Ge^{2+} part, the spectra in Figure 3.31 are deconvoluted into ten peaks, corresponding to $3d_{3/2}$ and $3d_{5/2}$ of Ge^0, Ge^{1+}, Ge^{2+}, Ge^{3+}, and Ge^{4+}, on the basis of the peak positions in reference [50]. The core-level shifts of the four oxidation states are 0.78 (Ge^{1+}), 1.57 (Ge^{2+}), 2.35 (Ge^{3+}), and 3.18 (Ge^{4+}), and the full width at half maximum of each spectrum is set as the same value of 1.4 eV. Here, the spectra of only Ge $3d_{5/2}$ photoelectrons are shown in Figure 3.33 after deconvolution from the original spectra. Clearly, in Figures 3.33(a) through (d), the integration area of Ge^{2+} decreases with increasing annealing temperature, while the Ge^0 and Ge^{4+} peaks grow. To further characterize the disproportionation reaction, the activation energy of the disproportionation reaction is calculated through XPS. First, we consider that the disproportionation reaction $2GeO \rightarrow GeO_2 + Ge$ follows the kinetic relationship:

$$I\left(Ge^{2+}\right) = I_0 f\left(t\right)\exp\left(-\frac{E_a}{K_B T}\right) \qquad (3\text{--}31)$$

where I (Ge^{2+}) is the integration area of Ge^{2+} after UHV annealing, I_0 is the initial Ge^{2+} integration area, f(t) is the pre-factor related to annealing time, E_a is the activation energy, and k_B is the Boltzmann constant. By plotting ln (I (Ge^{2+})) against 1/T, E_a of about 0.7 ± 0.2, eV is extracted from the slope. Moreover, for GeO disproportionation ($2GeO \rightarrow GeO_2 + Ge$), the Gibbs free energy per mole of reactant is calculated from room temperature (−150.62 kJ) to 700°C (−127.06 kJ).

Among the various experimental results reported for GeO_2/Ge interfaces, discussing the relation between GeO disproportionation and GeO desorption is helpful in understanding the various phenomena that take place at the interface of Ge-MOSFETs and providing a guideline for process optimization. Many groups reported that low temperature post deposition annealing treatment for a capped GeO_2/Ge is beneficial to

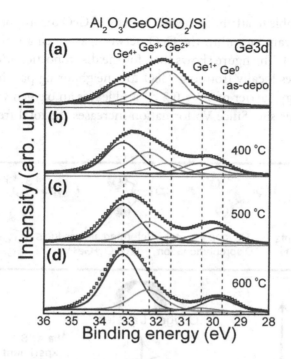

FIGURE 3.33 Peak deconvolution results for the Ge3d$_{5/2}$ spectra of the as-prepared Al$_2$O$_3$/GeO/SiO$_2$/Si stack and the stacks annealed at 400°C, 500°C, and 600°C for 30 minutes.

Source: Reproduced from S. K. Wang et al., "Kinetic study of GeO disproportionation into a GeO$_2$/Ge system using x-ray photoelectron spectroscopy", *Appl. Phys. Lett.* 101, 061907 (2012), with the permission of AIP Publishing.

improving the interface quality.[51, 52] For example, Lee et al. reported that by applying low-temperature oxygen annealing at around 400°C to the Ge-MOSFETs with a Y$_2$O$_3$/GeO$_2$/Ge stack, the sub-threshold slope value and the interface trap density were effectively reduced.[51] In that stack, it is difficult for GeO desorption to occur because of the blocking effect of the thick Y$_2$O$_3$ layer (15nm). Thus, the system could be reasonably regarded as a closed system. During GeO$_2$ formation, GeO (or Vo), which is regarded as the main contributor of interfacial traps, inevitably exists. In that case, we believe that after the 400°C annealing treatment, GeO disproportionation tends to occur together with the oxygen compensation and finally results in the annihilation of GeO and Vo. Therefore, it is quite reasonable to attribute the mechanism of GeO disproportionation to the oxygen transfer within the GeO network, as schematically proposed in Figure 3.34. The figure shows that, for the disproportionation reaction, oxygen moves locally from one GeO to a neighboring pair by overcoming the energy barrier of 0.7 ± 0.2 eV and leaves an oxygen vacancy (Vo) at its previous site. Since Vo formation increases the total free energy of

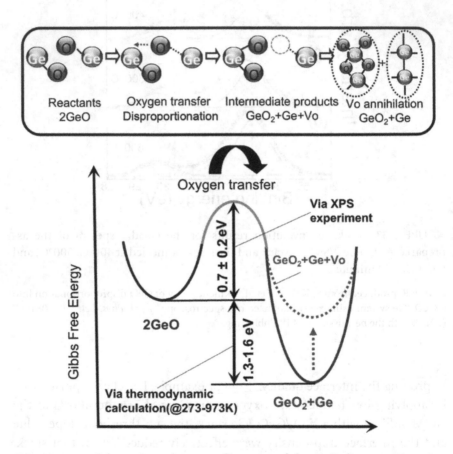

FIGURE 3.34 Schematic and energy diagrams of the GeO disproportionation mechanism. GeO disproportionation is attributed to oxygen transfer within the GeO network, where the activation energy extracted from Figure 3.33 is 0.7 ± 0.2 eV. Additionally, the free-energy differences are added in units of electron volts. Note that the x-axis has no physical meaning.

Source: Reproduced from S. K. Wang et al., "Kinetic study of GeO disproportionation into a GeO$_2$/Ge system using x-ray photoelectron spectroscopy", *Appl. Phys. Lett.* 101, 061907 (2012), with the permission of AIP Publishing.

the system, the reaction is inclined to follow the direction that annihilates the Vo. Since the reaction occurs in a closed system owing to the blocking effect of the capping layer, GeO desorption cannot occur; thus, Ge atoms have to recombine to annihilate Vo to further decrease the Gibbs free energy of the system. As a result, disproportionation finally reaches a stable state through the formation of Ge and GeO$_2$.

However, the mechanism of desorption of GeO from the GeO$_2$/Ge stack, which is attributed to the diffusion of the Vo generated in the redox reaction at the GeO$_2$/Ge interface, seems to contradict this assertion. For example, in the case of GeO$_2$/Ge without capping oxides, as demonstrated by Kita et al., the hysteresis in the C-V measurement becomes much larger after 600°C annealing treatment. On the basis of the photon absorption measurement, strong photon absorption at the GeO$_2$ sub-gap region was observed, which is attributed to the generation of Vo. In that case, the system is regarded as an open system because there is no capping layer, and the reaction is dominated by desorption. Once GeO desorption occurs, the desorbed GeO will no longer return to the system to act as the reactant. Therefore, on the basis of the aforementioned discussion, it is easy to understand the difference between these two reaction mechanisms by comparing the reaction conditions between GeO desorption from GeO$_2$/Ge and GeO disproportionation. For GeO desorption, it includes a diffusion process mediated by the Vo; incorporation of the Vo into GeO$_2$ increases the free energy of the GeO$_2$/Ge system, which lowers the energy barrier for the redox reaction that drives GeO desorption from the GeO$_2$ surface. Meanwhile, GeO disproportionation occurs in a closed system, under thermodynamic equilibrium, and the reaction strictly follows the direction that decreases the free energy of the system. Therefore, to improve the GeO$_2$/Ge interface quality for a high-performance Ge-MOSFET, constructing a closed system by capping GeO$_2$ with other dielectrics and taking account of GeO disproportionation is a good solution. (Author's note: This content also presents a good solution for forming solid GeO$_x$ or for making Ge nanocrystals by controlling the annealing temperature, which may act as the charge-trapping layer in memory devices.)

3.12 GEO DESORPTION-INDUCED ELECTRICAL DEGRADATION

We have clarified that O-vacancy incorporation into GeO$_2$ will result in GeO desorption and consequently lead to the degradation of GeO$_2$

bulk and GeO$_2$/Ge interface quality and the creation of sub-gap photo-adsorption and interfacial redox reaction. However, towards the application aspects of GeO$_2$/Ge stacks to Ge-MOS technology, it is also necessary to understand how GeO desorption affects the electrical properties of GeO$_2$/Ge stacks. Therefore, in this section, we concentrate on the electrical characterization of GeO$_2$/Ge stacks after GeO desorption.

From the thermodynamic point, increasing the O$_2$ partial pressure is one way to reduce O-vacancy concentration so as to enhance the quality of GeO$_2$.[47] Therefore, 22nm GeO$_2$ was formed by high-pressure oxidation (HPO) at 550°C in 70atm O$_2$ ambient for 25 minutes. After that, the GeO$_2$/Ge stacks received low temperature annealing at 400°C in O$_2$ ambient for one hour. Immediately thereafter, some GeO$_2$/Ge stacks were annealed in UHV at 500°C, for 1 minute. Then, Au electrodes were deposited on the GeO$_2$/Ge stacks with and without UHV by thermal evaporation.

To quantitatively estimate the distribution of D$_{it}$ at GeO$_2$/Ge interface, conductance method is used. Capacitance-voltage (C-V) and conductance-frequency (G-V) measurements are done at 200K on Ge MISCAPs with and without UHV annealing, as shown in Figure 3.35(a) through (d). Cooling down the sample to 200K helps prevent the minority carrier response. In Figure 3.35(a) and (c), although thermal minority–caused inversion response is suppressed, weak inversion response is still visible

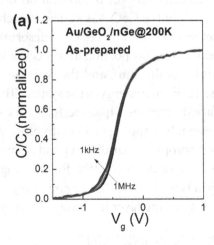

FIGURE 3.35 C-V (a) and G$_p$/ω ((b)) characteristics of the as-prepared Au/GeO$_2$/n-Ge MISCAP and the C-V (c) and G$_p$/ω ((d)) characteristics for the ones after UHVA at 500°C for one minute.

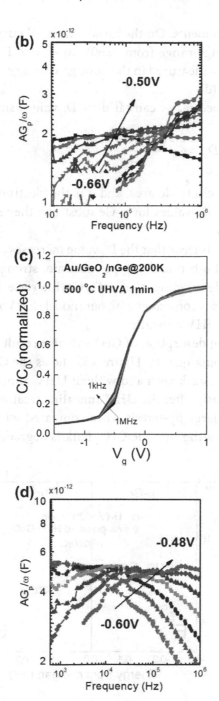

FIGURE 3.35 (Continued)

in our C-V measurements. On the basis of the C-V curves, we have identified the gate voltage range from -0.60V to -0.48V. Equivalent conductance (G_p/ω) spectra measured in the same gate voltage range, as shown in
Figure 3.35(b) and (d).

From the G-V spectra, we can calculate D_{it} value using:[53]

$$D_{it} = \frac{2.5\left(G_p/\omega\right)|max|}{qA} \tag{3-32}$$

where A is the Au electrode area, and q is the electron charge. We have
also calculated the D_{it} values for p-Ge substrate; the results are summarized in Figure 3.36.

In Figure 3.36, it is clear that the D_{it} value increases after UHV annealing treatment for both n and p type substrate, strongly indicating that
GeO desorption deteriorates the GeO₂/Ge interface by creating more
interface traps. This is consistent with our model that Vo is generated from
the GeO₂/Ge after UHV annealing.

Moreover, strong desorption of GeO will also result in the severe degradation of GeO₂ bulk quality. Figure 3.37 shows the C-V characteristics
of the Au/GeO₂/Ge stack with and without UHV annealing at 600°C for
30 seconds. Obviously, after the UHV annealing treatment, the C-V characteristic shows a large hysteresis when compared with the as-prepared
one, strongly suggesting that the GeO₂ bulk is degraded due to the GeO

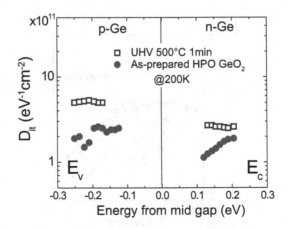

FIGURE 3.36   Energy distribution of the interface states density (D_{it}) estimated
by the conductance method at 200K.

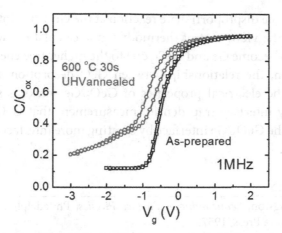

FIGURE 3.37 C-V characteristics of as-prepared Au/GeO₂/Ge MIS stack and the one with 600°C UHV annealing treatment for 30 seconds.

desorption. Due to the incorporation of Vo into the GeO₂ network, the residual Vo becomes the charge traps in GeO₂ and thus results in a large hysteresis during C-V measurement.

3.13 SUMMARY

In this chapter, the desorption kinetics of GeO from the GeO₂/Ge system have been investigated from various aspects. On the basis of the direct observation of Ge substrate consumption during GeO desorption, GeO has been confirmed to mainly desorb due to the reaction between GeO₂ and Ge. Moreover, we have concluded that the diffusion species is not the GeO molecule but oxygen vacancy by the ^{73}Ge and ^{18}O labeling technique. We conclude that the GeO desorption initiates from the GeO₂ surface in the uniform-desorption region. On the basis of the experimental results that O shows a much higher diffusion coefficient than Ge in GeO₂, the oxygen vacancy diffusion model has been proposed to explain the desorption mechanism of GeO from a GeO₂/Ge stack in the uniform-desorption region. The activation energy of about 2eV was obtained from the kinetic calculation. By using this model, we can successfully explain most of the experimental observations obtained so far. In addition, two kinds of GeO desorption (uniform and nonuniform) have been demonstrated, and the uniform one is likely to occur at relatively lower temperatures.

Moreover, the disproportionate reaction at the GeO$_2$/Ge interface is discussed from the viewpoint of thermodynamic calculation, where GeO is considered to become Ge and GeO$_2$ due to the higher free energy of GeO.

In addition, the relationship between GeO desorption and the degradation of the electrical properties of GeO$_2$/Ge stacks is studied. It is confirmed by interface trap density measurement that GeO desorption deteriorates the GeO$_2$/Ge interface by creating more interface traps within the band gap.

REFERENCES

1. R. H. Kingston, *Semiconductor Surface Physics*, Philadelphia: University of Pennsylvania Press, 1957.
2. A. J. Rosenberg, P. H. Robinson, and H. C. Gatos, "Thermal restoration of oxygenated germanium surfaces", *J. Appl. Phys.*, **29** (1958) 771.
3. D. Brennan, D. O. Hayward, and B. M. W. Trapnell, "Calorimetric determination of the heat of adsorption of oxygen on evaporated films of germanium and silicon", *J. Phys. Chem. Solids*, **14** (1960) 117.
4. J. J. Lander, and J. Morrison, "Structures of clean surfaces of germanium and silicon. I", *J. Appl. Phys.*, **34** (1963) 1403.
5. J. J. Lander and J. Morrision, "Low-energy electron-diffraction study of the surface reactions of germanium with oxygen and with iodine. II", *J. Appl. Phys.*, **34** (1963) 1411.
6. J. A. Dillon, Jr. and H. E. Farnsworth, "Work-function studies of germanium crystals cleaned by ion bombardment", *J. Appl. Phys.*, **28** (1957) 174.
7. S. P. Wolsky, "Contact potential measurements on graphite", *J. Appl. Phys.*, **29** (1958) 1132.
8. R. E. Schlier and H. E. Farnsworth, "Structure and adsorption characteristics of clean surfaces of germanium and silicon", *J. Chem. Phys.*, **30** (1959) 917.
9. H. D. Hagstrum, "Oxygen adsorption on silicon and germanium", *J. Appl. Phys.*, **32** (1961) 1020.
10. R. J. Madix and M. Boudart, "Sticking probabilities by an effusive beam technique: The germanium-oxygen system", *J. Catalysis.*, **7** (1967) 240.
11. J. B. Anderson and M. Boudart, "Sticking probability of oxygen molecules on single crystals of germanium", *J. Catalysis.*, **3** (1964) 216.
12. R. J. Madix and A. A. Susu, "Reactive scattering of atomic oxygen from clean elemental semiconductor surfaces", *Surf. Sci.*, **20** (1970) 377.
13. R. F. Lever and H. R. Wendt, "The reaction of oxygen with silicon and germanium at elevated temperatures by weight loss measurement", *Surf. Sci.*, **19** (1970) 435.
14. R. J. Madix, R. Parks, A. A. Susu, and J. A. Schwarz, "Chemical relaxation molecular beam studies of reactive gas-solid scattering: II. Reaction of ozone with heated germanium surfaces", *Surf. Sci.*, **24** (1973) 288.

15. A. A. Frantsuzov and N. I. Makrushin, "Temperature dependence of oxidation rate in clean Ge(111)", *Surf. Sci.*, **40** (1973) 320.
16. D. Schmeisser, R. D. Schnell, A. Bogen, F. J. Himpsel, D. Rieger, G. Landgren, and J. F. Morar, "Surface oxidation states of germanium", *Surf. Sci.*, **172** (1986) 455.
17. R. Ludeke, and A. Koma, "Oxidation of clean Ge and Si surfaces", *Phys. Rev. Lett.*, **34** (1975) 1170.
18. J. E. Rowe, "Photoemission and electron energy loss spectroscopy of GeO$_2$ and SiO$_2$", *Appl. Phys. Lett.*, **25** (1974) 576.
19. C. M. Garner, I. Lindau, J. N. Miller, P. Pianetta, and W. E. Spicer, "Growth and characterization of doped ZrO$_2$ and CeO$_2$ films deposited by bias sputtering", *J. Vac. Sci. Technol.*, **14** (1977) 372.
20. L. Surnev, "Oxygen adsorption on Ge(111) surface: I. Atomic clean surface", *Surf. Sci.*, **110** (1981) 439.
21. L. Surnev and M. Tikhov, "Oxygen adsorption on a Ge(100) surface: I. Clean surfaces", *Surf. Sci.*, **123** (1982) 505.
22. D. A. Hansen, and J. B. Hudson, "The adsorption kinetics of molecular oxygen and the desorption kinetics of GeO on Ge(100)", *Surf. Sci.*, **292** (1993) 17.
23. D. A. Hansen, and J. B. Hudson, "Oxygen scattering and initial chemisorption probability on Ge(100)", *Surf. Sci.*, **254** (1991) 222.
24. K. Prabhakaran and T. Ogino, "Oxidation of Ge (100) and Ge (111) surfaces: An UPS and XPS study", *Surf. Sci.*, **325** (1995) 263.
25. P. A. Redhead, "Thermal desorption of gases", *Vacuum*, **12** (1962) 203.
26. F. Jona, "Preparation of atomically clean surfaces of Si and Ge by heating in vacuum", *Appl. Phys. Lett.*, **6** (1965) 205.
27. D. A. Hansen, "Kinetics of the adsorption of O$_2$ and the desorption of GeO on Ge(100)", Ph.D. Dissertation, Rensselaer Polytechnic Institute, 1990, p. 206.
28. K. Kita, S. Suzuki, H. Nomura, T. Takahashi, T. Nishimura, and A. Toriumi, "Direct evidence of GeO volatilization from GeO$_2$/Ge and impact of its suppression on GeO$_2$/Ge metal-insulator-semiconductor characteristics", *Jpn. J. Appl. Phys.*, **47** (2008) 2349.
29. K. Kita, M. Yoshida, T. Nishimura, K. Nagashio, and A. Toriumi, "Spectroscopic Ellipsometry Study on Defects Generation in GeO$_2$/Ge stacks", *Int. Conf. on Solid State Devices and Materials*, 2009, p. 1008.
30. Y.-K. Sun, D. J. Bonser, and T. Engel, "Spatial inhomogeneity and void-growth kinetics in the decomposition of ultrathin oxide overlayers on Si(100)", *Phys. Rev. B*, **43** (1991)14309.
31. Y.-K. Sun, D. J. Bonser, and T. Engel, "Thermal decomposition of ultrathin oxide layers on Si(100)", *J. Vac. Sci. Technol. A*, **10** (1992) 2314.
32. Y. Kobayashi and K. Sugii, "Thermal decomposition of very thin oxide layers on Si(111)", *J. Vac. Sci. Technol. A*, **10** (1992) 2308.
33. R. Tromp, G. W. Rubloff, P. Balk, F. K. LeGoues, and E. J. van Loenen, "High-temperature SiO$_2$ decomposition at the SiO$_2$/Si interface", *Phys. Rev. Lett.*, **55** (1985) 2332.

34. M. Liehr, J. E. Lewis, and G. W. Rubloff, "Kinetics of high-temperature thermal decomposition of SiO_2 on Si(100)", *J. Vac. Sci. Technol. A.*, **5** (1987) 1559.
35. P. Viscor, "In-situ measurements of the density of amorphous germanium prepared in ultra-high vacuum", *J. Non-Cryst. Solids*, **101** (1988) 170.
36. E. Huger, U. Tietze, D. Lott, H. Bracht, D. Bougeard, E. E. Haller, and H. Schmidt, "Self-diffusion in germanium isotope multilayers at low temperatures", *Appl. Phys. Lett.*, **93** (2008) 162104.
37. J. C. Mikkelsen, Jr, "Self-diffusivity of network oxygen in vitreous SiO_2", *Appl. Phys. Lett.*, **45** (1984) 1187.
38. C. W. Magee and R. E. Honig, "Depth profiling by SIMS- depth resolution, dynamic range and sensitivity", *Surf. Interface Anal.*, **4** (1982) 35.
39. J. Crank, *The Mathematics of Diffusion*, Oxford: Clarendon Press, 1975, 2nd ed., p. 14.
40. H. Hosono, Y. Abe, D. L. Kinser, R. A. Weeks, K. Muta, and H. Kawazoe, "Nature and origin of the 5-eV band in SiO_2:GeO_2 glasses", *Phys. Rev. B*, **45** (1992) 11445.
41. M. Takahashi, K. Ichii, Y. Tokuda, T. Uchino, T. Yoko, J. Nishi, and T. Fujiwara, "Photochemical reaction of divalent-germanium center in germanosilicate glasses under intense near-ultraviolet laser excitation", *J. Appl. Phys.*, **92** (2002) 3442.
42. T. Sasada, Y. Nagakita, M. Takenaka, and S. Takagi, "Surface orientation dependence of interface properties of GeO_2/Ge metal-oxide-semiconductor structures fabricated by thermal oxidation", *J. Appl. Phys.*, **106** (2009) 073716.
43. HSC Chemistry® 6.0 database. (OUTOKUMPU Technology).
44. K. Kita, S. K. Wang, M. Yoshida, C. H. Lee, K. Nagashio, T. Nishimura, and A. Toriumi, "Comprehensive study of GeO_2 oxidation, GeO desorption and GeO_2-metal interaction, -understanding of Ge processing kinetics for perfect interface control-", *Tech. Dig.—Int. Electron Devices Meet.*, 2009, p. 693.
45. A. Schacht, C. Sternemann, A. Hohl, H. Sternemann, Ch. J. Sahle, M. Paulus, and M. Tolan, "Temperature-induced obliteration of sub-oxide interfaces in amorphous GeO", *J. Non-Cryst. Solids*, **355** (2009) 1285.
46. C. J. Sahle, C. Sternemann, H. Conrad, A. Herdt, O. M. Feroughi, M. Tolan, A. Hohl, R. Wagner, D. Lützenkirchen—Hecht, R. Frahm, A. Sakko, and K. Hämäläinen, "Phase separation and nanocrystal formation in GeO", *Appl. Phys. Lett.*, **95** (2009) 021910.
47. S. K. Wang, K. Kita, C. H. Lee, T. Tabata, T. Nishimura, K. Nagashio, and A. Toriumi, "Desorption kinetics of GeO from GeO_2/Ge structure", *J. Appl. Phys.* **108** (2010) 054104.
48. T. Hosoi, K. Kutsuki, G. Okamoto, A. Yoshigoe, Y. Teraoka, T. Shimura, and H. Watanabe, "Synchrotron radiation photoemission study of Ge_3N_4/Ge structures formed by plasma nitridation", *Jpn. J. Appl. Phys.*, **50** (2011) 10PE03.
49. K. Kato, S. Kyogoku, M. Sakashita, W. Takeuchi, H. Kondo, S. Takeuchi, O. Nakatsuka, and S. Zaima, "Control of interfacial properties of Al_2O_3/Ge gate stack structure using radical nitridation technique", *Jpn. J. Appl. Phys.*, **50** (2011) 10PE02.

50. K. Vijayarangamuths, S. Rath, D. Kabiraj, D. K. Avasthi, P. K. Kulriya, V. N. Singh, and B. R. Mehta, "Ge nanocrystals embedded in a matrix formed by thermally annealing of Ge oxide films", *J. Vac. Sci. Technol. A*, **27** (2009) 731.

51. C. H. Lee, T. Nishimura, T. Tabata, S. K. Wang, K. Nagashio, K. Kita, and A. Toriumi, "Ge MOSFETs Performance: Impact of Ge Interface Passivation", *Tech. Dig.-Int. Electron Devices Meet*, (2010) 416.

52. R. Zhang, P. C. Huang, N. Taoka, M. Takenaka, and S. Takagi, "High mobility Ge pMOSFETs with 0.7 nm ultrathin EOT using HfO$_2$/Al$_2$O$_3$/GeO$_x$/Ge gate stacks fabricated by plasma post oxidation", *Dig. Tech. Pap.-Symp. VLSI Technol*, (2012) 161.

53. E. H. Nicollian, and A. Goetzberger, "The Si-SiO$_2$ interface electrical properties as determined by the metal-insulator-silicone conductance technique", *Bell Syst. Tech. J*, **46** (1967) 1055.

CHAPTER 4

Structural Transition Kinetics in GeO$_2$/Ge

4.1 INTRODUCTION

In Ge-based MOSFETs, the GeO$_2$/Ge stack is as fundamental a structure as SiO$_2$/Si is in Si-based MOSFETs. However, compared to SiO$_2$/Si, many things need to be reconsidered due to the thermal instability of the GeO$_2$/Ge interface in GeO$_2$/Ge stacks: e.g., GeO desorption and sub-gap states formation.[1, 2] Therefore, depositing a structurally reliable and high quality GeO$_2$ on Ge comparable to SiO$_2$ on Si turns out to be an important issue.

Amorphous GeO$_2$ is considered a chemical and structural analogue of SiO$_2$. Due to a high energy barrier for crystallization, an amorphous network of SiO$_2$ does not show long-range ordering, even after receiving high-temperature treatment on Si substrate.[3] However, compared to SiO$_2$, amorphous GeO$_2$ has a broader distribution in the Ge-O-Ge angle;[4] it is expected that GeO$_2$ on Ge behaves differently from SiO$_2$ on Si after thermal treatment.

In this chapter, we report on the crystallization of GeO$_2$ on Ge and discuss how it is strongly related to the Ge/GeO$_2$ system. We infer that the energy barrier for crystallization is reduced by the interface reaction between GeO$_2$ and Ge, possibly owing to the introduction of oxygen vacancy into the GeO$_2$ network. Moreover, this chapter will also present a

DOI: 10.1201/9781003284802-4

model to explain the origin of voids formation in the nonuniform desorption region.

4.2 STRUCTURES OF GERMANIUM OXIDES

4.2.1 Structure of GeO_2

Germanium dioxide (GeO_2), appears as white powder or colorless crystals and has structure forms that are commonly hexagonal, tetragonal, and amorphous. It always forms a native passivation layer on pure germanium when in contact with oxygen. Its melting point is about 1100°C, and it can dissolve in both alkali and acid solutions, and certain structure forms can also dissolve in water. It is thermally and chemically unstable and does not quite suit dielectric material, which is one of the main disadvantages compared to silicon.[5]

The structure of GeO_2 has generally been considered to be comparable to that of silica glass despite differences in bond lengths, angles, and the relative size of Ge versus Si. At room temperature, GeO_2 has four phases: amorphous, α-quartz-like crystalline GeO_2, cristobalite-like, and rutile-like crystalline. The structure of crystalline α-quartz-like GeO_2 and rutile-like GeO_2 are shown in Figures 4.1(a) and (b).

α-quartz-like GeO_2 has a ($P3_221$) hexagonal structure and a rutile-like tetragonal ($P4_2$/mnm) structure.[6, 7] The α-quartz-like GeO_2 has been shown to be stable at the high-temperature phase,[8] and, while the structure is very similar to that of α-quartz, there are some distinct differences. As shown in Figure 4.1(a), α-quartz-like GeO_2 has a unit of GeO_4 tetrahedron. The GeO_4 tetrahedra are more distorted due to greater variation in the O-Ge-O angles (106.3~113.1°) within the tetrahedron, with a Ge-O-Ge angle of 130.1°. However, in α-quartz, the O-Si-O angles within the SiO_4 are more uniform (108.3~ 110.7°), with an Si-O-Si angle of 144.0°.[9]

Rutile GeO_2 is the stable phase at room temperature, which transforms to the α-quartz-like GeO_2 structure at 1281K.[8] The rutile GeO_2 polymorph has a structure similar to that of stishovite[7] (as depicted in Figure 4.1(b)), and the two axial bonds within the GeO_6 polyhedron (1.902 ± 0.001 Å) are longer than the four equatorial Ge-O bonds (1.872 ± 0.001 Å). Conversely, the two independent Ge-O distances in the α-quartz-like GeO_2 structure are similar at 1.737 ± 0.003 Å and 1.741 ± 0.002 Å.[6] (Author's note: Recently, rutile GeO_2 has been considered for future application in power devices because of its ultrawide-band-gap property, 4.68eV, and because it can be ambipolarly doped. Some initial work has been done by Kioupakis

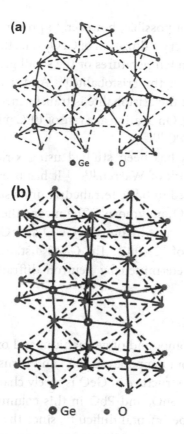

FIGURE 4.1 The structure of (a) crystalline α-quartz-like GeO$_2$ and (b) rutile-like GeO$_2$.

et al. (*APL* 114(10):102104, 2019). In the author's opinion, high-quality growth rutile GeO$_2$ is quite possible if the motion of oxygen vacancy is carefully controlled.)

Cristobalite has a deceptively simple crystal structure. The Ge atoms are arranged in the same way that carbon atoms are arranged in the cubic diamond structure, which means that each Ge atom has four neighboring silicon atoms in a perfect tetrahedral arrangement. The oxygen atoms are located halfway along the vector between nearest-neighbor Ge atoms. The cristobalite-like GeO$_2$ was first observed by H. Bohm in 1968.[10] The experiments on the crystallization of GeO$_2$ glass proved that a third crystalline form can be obtained by long-time heat treatment at about 600°C in bulk samples. The new GeO$_2$ modification is obviously isostructural to cristobalite. It was not possible, however, to prepare this modification in pure form; additional crystallization of GeO_2-quartz was always observed. Higher crystallization temperatures or the small grain size of the sample prevented formation of the cristobalite.[10] And it has been reported by Yamaguchi et al. that the cristobalite is transformed into α-quartz in the later stages of heating. On heating at a rate of 10°C min⁻¹, only the α-quartz crystallizes above 810°C.[11]

Amorphous GeO_2 has been studied using x-ray diffraction (XRD) in the pioneering work of Warren.[12, 13] It has been confirmed that the Ge atoms are arranged in basic tetrahedral units such as those found in the α-quartz-like GeO_2 polymorph. The means the Ge-O-Ge intertetrahedral angle was estimated from the Ge-O and Ge-Ge distances to be 130.1°, with a range of 121~147°. The Ge-O distances in the amorphous GeO_2 structure is determined by neutron diffraction and anomalous x-ray scattering.[14]

4.2.2 Structure of GeO

GeO is a chemical compound of germanium and oxygen. It appears as a yellow powder at room temperature (RT) and turns brown on heating at 650°C. However, the structure of GeO is rarely characterized when compared with CO, SiO, SnO, and PbO in this column. This appears to be partly because of experimental difficulty, since the vapors of hot germanium dioxide, which are normally used to get gaseous monoxide diatoms as dissociate ion product, react sharply with the quartz components of discharge tubes, etc.[15] Majumdar and Mohan[16] have reported a well-developed absorption spectrum (D system) of this molecule. Extensive data were presented by these authors, along with the expected isotopic shifts in many of the bands. On the basis of a general discussion on the trend and behavior of the various excited states of homologous monoxides, viz. CO, SiO, SnO, and PbO, it was predicted that one should also identify a state B for GeO, corresponding to the B state of SnO and PbO. Tewari and Mohan[16] have conducted a measurement of the thermal emission spectrum of GeO, and they found that the B system of GeO should bear a close resemblance to the already-known B system of SnO. From the viewpoint of electrochemical stability, it was found that solid GeO has its own region of stability in the electrochemical stability diagram,[17] similar to SnO but unlike unstable SiO. Lin et al.[18] have calculated the stable crystalline structure of GeO using *Ab initio* methods, trying the tetragonal

PbO structure, the rhombohedral GeTe structure, and the orthorhombic GeS structure. They found that that crystalline GeO and SiO are both stable in the orthorhombic structure (similar to orthorhombic GeS, which consists of threefold puckered sheets of coordinated Ge and S atoms with 95° bond angles) but with planar oxygen sites. In these ten-electron systems, formally, one electron is transferred from the O to the Ge to give each atom five valence electrons. Ge then makes three bonds at near 90° bond angles with its p electrons. The O atom is now isoelectronic to N in Si$_3$N$_4$; it forms three planar bonds at 120° with an sp^2 configuration and keeps two p electrons in a lone p-π pair. This is consistent with Binder et al.,[19] who found that a random network of GeO consists mainly of threefold Ge and O sites. The orthorhombic GeO (GeS) structure is schematically shown in Figure 4.2.

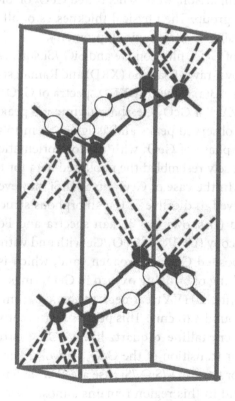

FIGURE 4.2 Orthorhombic GeO structure. Green/darker balls = Ge, red/lighter balls = O. Dashed lines are weak Ge-Ge bonds.[16]

4.3 STRUCTURAL TRANSITION IN GEO$_2$/GE SYSTEMS

In this subsection, we discuss the crystallization of GeO$_2$ in GeO$_2$/Ge systems upon annealing in UHV condition.

Prior to deposition, Ge (100) substrates were pre-treated by using ultrasonic cleaner—one minute in 25% HCl solution, 30 seconds in H$_2$O$_2$/ammonia/H$_2$O (1:0.5:100) solution—followed by an etch treatment in 5% HF solution. Immediately after, the substrates were rinsed in de-ionized water and blown dry in nitrogen. Subsequently, 140~540nm GeO$_2$ films were deposited on Ge (100) by sputtering in an oxygen-argon mixture, followed by ultra-high vacuum annealing (UHVA) at about 660°C for one minute. To confirm the stoichiometry of the as-sputtered GeO$_2$ film, x-ray photoelectron spectroscopy (XPS) measurement was performed.

The background vacuum level of the main chamber was about 8×10^{-8}~5×10^{-7}Pa, which can be reasonably regarded as an oxygen-free condition. For comparison, we also deposited GeO$_2$ on SiO$_2$ (1um)/Si substrates as control group. The physical thicknesses of all the GeO$_2$ layers were confirmed using spectroscopic ellipsometry.

The structure of GeO$_2$ films on Ge and SiO$_2$/Si substrates after UHVA was investigated by x-ray diffraction (XRD) and Raman spectroscopy measurements. Figure 4.3(a) shows the XRD spectra of GeO$_2$ on Ge (100) and SiO$_2$/Si after UHVA. For GeO$_2$/Ge, sharp diffraction peaks were observed. We identified all observed peaks and assigned them to the formation of the α-quartz-like phase of GeO$_2$ with random orientation because these peak positions closely resembled the published data for the α-quartz-like phase of GeO$_2$.[6] In the case of GeO$_2$ on SiO$_2$/Si, however, no diffraction peaks were observed, indicative of the amorphous structure of the film. Figure 4.3(b) and (c) shows the Raman spectra and Fourier transform infrared spectroscopy (FT-IR) of GeO$_2$/Ge with and without UHVA.

For the as-deposited GeO$_2$, a broaden band, which is assigned to the symmetric stretching of bridging oxygen in GeO$_4$ rings, is observed from 370 to 500 cm^{-1}. After UHVA treatment, the band becomes much sharper, with a peak at around 440 cm^{-1}. This peak is consistent with the A$_1$ symmetric mode in crystalline α-quartz-like GeO$_2$,[20] strongly indicating the disorder-order transition of the GeO$_2$ network. For comparison, the Raman spectra for the GeO$_2$/SiO$_2$/Si case are also shown in the inset. It is clear that the band in this region remains almost the same after UHVA treatment, indicating that without the existence of Ge substrate, the structural ordering transition in GeO$_2$ does not occur. This result is confirmed

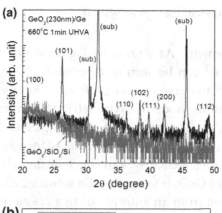

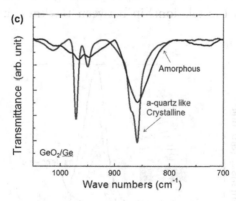

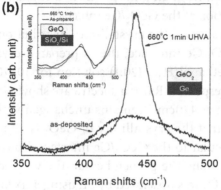

FIGURE 4.3 (a) XRD spectra of the GeO_2 on Ge (100) and SiO_2/Si after UHVA at 660°C for one minute. Sharp diffraction peaks can be attributed to the α-quartz-like phase of GeO_2. The corresponding peak labels are displayed as well. (b) Raman spectra of the as-deposited GeO_2/Ge and the one annealed at 660°C for one minute. Raman spectra of the as-prepared GeO_2/SiO_2/Si and the one annealed at 660°C for one minute are shown in the inset. (c) FT-IR spectra of the as-deposited GeO_2/Ge and the one annealed at 660°C for one minute.

Source: Reproduced from S. K. Wang et al., "Kinetic Effects of O-Vacancy Generated by GeO_2/Ge Interfacial Reaction", *Japanese Journal of Applied Physics* 50 (2011) 10PE04, with the permission of the Japan Society of Applied Physics.

by FT-IR measurement. As shown in Figure 4.3(c), broaden peaks at 857cm^{-1} and 971cm^{-1} can be seen in amorphous GeO$_2$. These peaks are attributed to the transverse optical (TO) vibration mode and longitude optical (LO) vibration mode of asymmetric stretching modes of bridging O. After UHVA treatment, both peaks assigned to TO and LO modes become sharper, indicating that the variation of the vibration of the bridging O becomes smaller. Hence, we can conclude unambiguously that crystalline α-quartz-like GeO$_2$ is formed on Ge substrate after UHVA, and this structural transition from an amorphous to a crystalline phase is closely related to the reaction at the GeO$_2$/Ge interface.

To further characterize the crystallinity distribution of GeO$_2$ perpendicular to the GeO$_2$/Ge interface, the as-prepared crystalline GeO$_2$ was etched using a methanol/H$_2$O (20:1) solution for 10, 70, and 80 minutes successively. The respective Raman spectra are shown in Figure 4.4.

The band width at 440cm^{-1} remains unchanged with the reduction of GeO$_2$, indicating that the crystallinity of GeO$_2$ is almost uniformly distributed perpendicular to the GeO$_2$/Ge interface. To confirm the uniformity of the GeO$_2$ along the perpendicular, the water etching rate of the UHVA-treated GeO$_2$ was measured by SE using a fixed n and k value set. Figure 4.5 shows the relationship between thickness and etching time

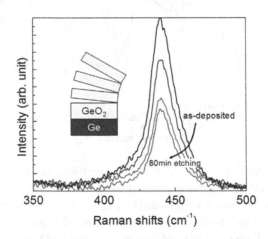

FIGURE 4.4 Raman spectra of as-annealed GeO$_2$/Ge sample with water etching for 0, 10, 70, and 80 minutes. The band widths are almost the same, suggesting that GeO$_2$ is uniformly crystallized perpendicular to the GeO$_2$/Ge interface. The schematic of this experiment is also shown in the inset for reference.

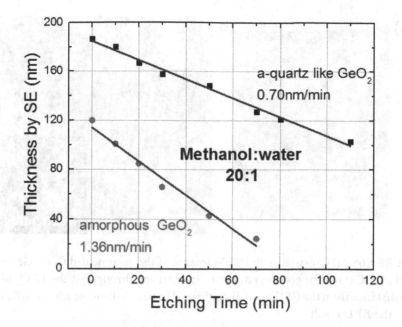

FIGURE 4.5 Relationship between thickness and etching time for UHVA-treated and as-sputtered amorphous GeO$_2$/Ge sample. The etching rate of α-quartz-like GeO$_2$ (0.70nm/min) is much slower than that of amorphous GeO$_2$ (1.35nm/min). The linear relationship indicates the film with UHVA treatment has a good uniformity along the perpendicular.

for a UHVA-treated GeO$_2$/Ge sample. The as-sputtered GeO$_2$/Ge is also included as a control group. The etching rates extracted from the slope indicate that crystalline GeO$_2$ shows more water inertness than amorphous GeO$_2$. Moreover, the linear relationship between thickness and etching time suggests the uniformity of the crystallization perpendicular to the film surface.

This conclusion is also directly confirmed by the cross-sectional transmission electron microscopy (TEM) image, as depicted in Figure 4.6(a) and (b), in which crystalline morphology is observed throughout the GeO$_2$ film.

Moreover, in Figure 4.6(b), the lattice constants determined by XRD are consistent with the distance between neighboring Ge atoms and the electron diffraction result from the (101) plane, as depicted in the inset of Figure 4.6(b). Therefore, we conclude that the crystallization of GeO$_2$ in GeO$_2$/Ge is not localized at the interface but is present throughout the

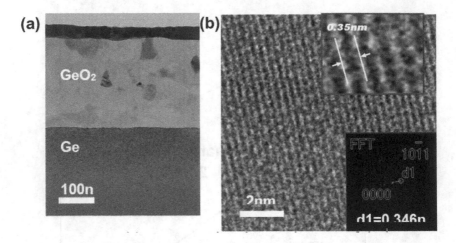

FIGURE 4.6 (a) Cross-sectional TEM images of the as-annealed GeO_2/Ge (100) stack. (b) GeO_2 with good crystallinity is observed throughout the GeO_2 film. The distance from the (101) plane is confirmed to be 0.364nm, which is consistent with the XRD result.

GeO_2 film. The crystallization of GeO_2 in GeO_2/Ge is quite different from the observations in the SiO_2/Si case.[3, 21]

In the field of glass or ceramic, the creation of microcrystalline within the glass network is commonly referred as "devitrification reaction". Causes of devitrification, commonly referred to as "devit", can include holding a high temperature for too long, which causes the nucleation of crystals, or the presence of foreign residue such as dust on the surface of the glass or inside the kiln prior to firing can be a key nucleation point where crystals can propagate easily. In the case of GeO_2 devitrification, since there is no observation of GeO_2 crystallization on SiO_2, the over-annealing case can be eliminated. Therefore, the crystallization of GeO_2 on Ge is attributed to the presence of foreign residue.

In chapter 3, an oxygen vacancy model was proposed in which the GeO_2/Ge interface was considered as the source for oxygen vacancy generation. Concerning the crystallization of GeO_2, therefore, we infer that the crystallization of α-quartz-like GeO_2 is also associated with the incorporation of oxygen vacancy into the GeO_2 network. As demonstrated in chapter 3, the diffusion coefficients of Ge are much smaller than O in GeO_2; therefore, in amorphous GeO_2, Ge atoms are arranged in basic tetrahedral units such as those found in the α-quartz-like GeO_2 polymorph; the crystallization is mainly achieved by the arrangement of oxygen atoms

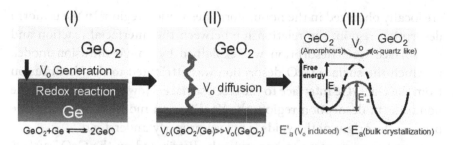

FIGURE 4.7 Schematic of oxygen vacancy-induced GeO$_2$ crystallization model. Note E$_a$ is the activation energy for bulk GeO$_2$ crystallization, and E'$_a$ is the activation energy for GeO$_2$ crystallization in GeO$_2$/Ge stacks.

in the GeO$_2$ network. For the crystallization of GeO$_2$ itself, since the equibrillium concentration of oxygen vacancy in GeO$_2$ is quite low even at a high temperature, oxygen atoms have very little freedom to move within the GeO$_2$ network. The crystallization requires much energy for the rearrangement of the oxygen atoms. In the case of GeO$_2$/Ge, however, we think large amounts of oxygen are diffused from the interfacial region. Under oxygen vacancy, the oxygen atoms in the GeO$_2$ network may have more freedom for rearrangement, resulting in lower crystallization activation energy. Here, we have to point out that although GeO$_2$ is crystallized throughout the film, the initial nuclei are provided at the interface. With the help of oxygen vacancy, the interfacial crystalline GeO$_2$ may also serve as a template to reduce the surface energy for GeO$_2$ at the non-interfacial region. The schematics of the oxygen vacancy–induced crystallization model are proposed in Figure 4.7. A similar vacancy-assisted crystallization model has been proposed in the study of Si crystallization elsewhere.[22]

In conclusion, the amorphous-to-crystalline α-quartz-like-GeO$_2$ transition has been observed after UHVA treatment. An oxygen vacancy-induced crystallization model with relatively lower activation energy has been proposed. Controlling the oxygen vacancy injection from the interfacial reaction should be the key to controlling the crystallization of GeO$_2$ in the GeO$_2$/Ge stack.

4.4 ORIGIN OF THE VOIDS IN NONUNIFORM DESORPTION REGION

In Chapter 3, for GeO desorption from GeO$_2$/Ge, desorption kinetics are discussed from the viewpoint of a uniform-nonuniform one, where voids are locally observed in the nonuniform desorption region. In the uniform desorption region, the relationship between the interfacial reaction and the surface GeO desorption was explained by an Vo diffusion model, in which the surface GeO desorption was attributed to the Vo diffusion from the GeO_2/Ge interface to the GeO_2 surface. However, concerning the nonuniform desorption region, the Vo diffusion model is not applicable because the origin of the voids is not yet properly understood.

Since the voids seem to be randomly distributed on the GeO_2 surface in the nonuniform desorption region, we infer that the voids' formation might be closely related to the crystallization of GeO_2 by considering the similarity to the random nucleation observed generally in the crystallization.

Figures 4.8(a) and (b) show the surface and cross-sectional atomic force microscopy (AFM) images of the as-crystallized 540nm GeO_2/Ge stacks. In Figure 4.8(a), a crystalline-grain-like morphology with clear boundaries is observed, while Figure 4.8(b) shows the cross-sectional profile of the boundary region as marked in Figure 4.8(a). In the "grain boundary" region, a valley with a depth over 70nm is observed. On the basis of these results, together with our previous observation about the nonuniform desorption of GeO with locally formed voids, we infer that

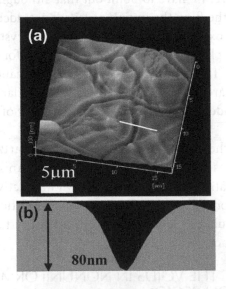

FIGURE 4.8   (a) AFM image of 540nm GeO_2/Ge stack annealed in UHV at 660°C for one minute. (b) Cross-sectional AFM image of the boundary region is marked in Figure 4.9(a).

these two effects are closely related to each other. Vo incorporated into the GeO$_2$ network may form the ordered structure locally. Vo tends to further accumulate at defective boundary sites with crystallization because the Vo diffusion barrier at the grain boundaries is likely to be much lower than that in the ordered GeO$_2$ network. As a result, the GeO desorption rate increases significantly in the boundary regions.

4.5 UNIFIED MODEL FOR KINETIC EFFECTS IN GEO$_2$/GE SYSTEMS

In this subsection, a unified model considering the incorporation of V$_O$ from the interfacial reaction is proposed. Figure 4.9 schematically

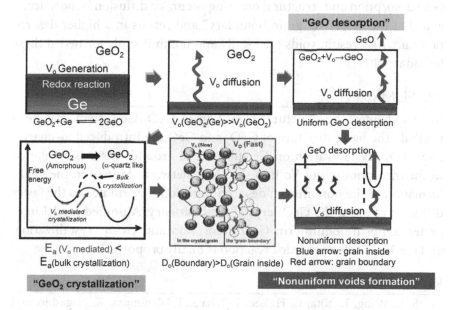

FIGURE 4.9 Schematic of interfacial reaction in GeO$_2$/Ge- and Vo-induced kinetic effects in GeO$_2$/Ge stacks. Vo generated from the interface diffuses into GeO$_2$ bulk. On the one hand, it provides a new route for the O rearrangement in GeO$_2$ with lower activation energy. On the other hand, it also results in the GeO desorption from the O-deficient GeO$_2$ surface. When both GeO desorption and structural ordering occurs, Vo diffusion is accelerated at the grain boundary and results in a higher desorption rate. As a result, voids are locally and nonuniformly formed at these boundary sites.

Source: Reproduced from S. K. Wang et al., "Kinetic Effects of O-Vacancy Generated by GeO$_2$/Ge Interfacial Reaction", *Japanese Journal of Applied Physics* 50 (2011) 10PE04, with the permission of the Japan Society of Applied Physics.

summarizes the impacts of interfacial reaction on GeO_2/Ge stack restructuring.

By considering the V_O diffusion in GeO_2 film, most of the experimental results observed so far can be reasonably explained. Vo is first generated at the GeO_2/Ge interface through the redox reaction. The gradient of Vo concentration between the interfacial region and the GeO_2 bulk, results in a diffusion flux of Vo from the interface to the GeO_2 surface. On the one hand, when Vo diffuses to the GeO_2 surface, the GeO_2 surface becomes O-deficient, and GeO desorption occurs from the GeO_2 surface. On the other hand, Vo incorporation into GeO_2 bulk provides a new route for the O atom to rearrange GeO_2 film to the ordered structure with lower activation energy. It is α-quartz-like GeO_2. And at a certain stage, when both GeO desorption and structural ordering occur, Vo diffusion is more accelerated at the defective "grain boundary" and results in a higher desorption rate. As a result, voids are locally and nonuniformly formed at these boundary sites.

4.6 SUMMARY

In this chapter, the structural transition in GeO_2/Ge systems has been studied. The basic structure of GeO_2 and GeO are introduced. α-quartz-like GeO_2 crystallization on Ge substrate at around 660°C is attributed to the incorporation of Vo into the GeO_2 network. Moreover, the voids' formation in the nonuniform desorption region is attributed to the faster diffusion of Vo at the GeO_2 crystalline boundary. A unified model integrates uniform/nonuniform GeO desorption and GeO_2 crystallization, and GeO_2/Ge interface redox reaction is finally proposed.

REFERENCES

1. S. K. Wang, K. Kita, C. H. Lee, T. Tabata, T. Nishimura, K. Nagashio, and A. Toriumi, "Desorption kinetics of GeO from GeO_2/Ge structure", *J. Appl. Phys.*, **108** (2010) 054104.
2. K. Kita, S. Suzuki, H. Nomura, T. Takahashi, T. Nishimura and A. Toriumi, "Direct evidence of GeO volatilization from GeO_2/Ge and impact of its suppression on GeO_2/Ge metal-insulator-semiconductor characteristics", *Jpn. J. Appl. Phys.*, **47** (2008) 2349.
3. A. Munkholm, S. Brennan, F. Comin, and L. Ortega, "Observation of a distributed epitaxial oxide in thermally grown SiO_2 on Si (001)", *Phys. Rev. Lett.*, **75** (1995) 4254.
4. J. A. E. Desa, A. C. Wright, and R. Sinclair, "A neutron diffraction investigation of the structure of vitreous Germania", *J. Non-Cryst. Solids*, **99** (1988) 276.

5. D. R. Lide, *CRC Handbook of Chemistry and Physics*, Boca Raton, FL: CRC Press, 2005, 86th ed.
6. G. S. Smith, and P. B. Isaacs, "The crystal structure of quartz-like GeO$_2$", *Acta Crystallographica*, **17** (1964) 842.
7. W. H. Baur, and A. A. Khan, "Rutile-type compounds. IV. SiO$_2$, GeO$_2$ and a comparison with other rutile-type structure", *Acta Crist.*, **B27** (1971) 2133.
8. A. W. Laubengayer, and D. S. Morton, "Germanium. XXXIX. The polymorphism of germanium dioxide", *J. Am. Ceram. Soc.*, **49** (1932) 2302.
9. J. D. Jorensen, "Compression mechanisms in α-quartz structures-SiO$_2$ and GeO$_2$", *J. Appl. Phys.*, **49** (1978) 5473.
10. H. Bohm, "The cristobalite modification of GeO$_2$", *Kurze Originalmitteilungen*, **55** (1968) 649.
11. O. Yamaguchi, K. Kotera, M. Asano, and K. Shimizu, "Crystallization and Transformation of amorphous GeO$_2$ derived from hydrolysis of germanium Isopropoxide", *J. Chem. Soc. Dalton Trans.* (1982) 1907.
12. B. E. Warren, "The diffraction of X-rays in glass", *Phys. Rev.*, **45** (1934) 657.
13. B. E. Warren, "X-ray determination of the structure of glass", *J. Am. Ceram. Soc.*, **17** (1934) 249.
14. D. L. Price, M.-L. Saboungi, and A. C. Barnes, "Structure of vitreous Germania", *Phys. Rev. Lett.*, **81** (1998) 3207.
15. D. P. Tewari, and H. Mohan, "Band spectrum of the molecule germanium monoxide (GeO) in the visible region", *J. Mol. Spec.*, **39** (1971) 290.
16. K. Majumdar, and H. Mohan, "Ultraviolet absorption spectrum of germanium monoxide molecule", *Indian J. Pure App. Phys.*, **6** (1969) 183.
17. M. Pourbaix, *Atlas of Electrochemical Equilibria in Aqueous Solutions*, Houston: National Association of Corrosion Engineers, 1966.
18. L. Lin, K. Xiong, and J. Robertson, "Atomic structure, electronic structure, and band offsets at Ge: GeO: GeO$_2$ interfaces", *Appl. Phys. Lett.*, **97** (2010) 242902.
19. J. F. Binder, P. Brovquist, and A. Pasquarello, presented at EMRS, June 2009.
20. J. F. Scott, "Raman spectra of GeO$_2$", *Phys. Rev. B*, **1** (1970) 3488.
21. I. Takahashi, T. Shimura, and J. Harada, "Structure of silicon oxide on Si(001) grown at low temperatures", *J. Phys. Condens. Matter*, **5** (1993) 6525.
22. J. Nakata, "Mechanism of low-temperature ($\leq$300°C) crystallization and amorphization for the amorphous Si layer on the crystalline Si substrate by high-energy heavy-ion beam irradiation", *Phys. Rev. B.*, **43** (1991) 14643.

CHAPTER 5

Oxidation in GeO$_2$/ Ge Stacks

5.1 INTRODUCTION

Oxidation reaction on semiconductors is an essential issue in the processing of microelectronic devices and attracts many groups to the study on silicon oxidation.[1, 2] For silicon, perfect gate dielectric formation and post-deposition annealing can be achieved by thermal oxidation treatment. However, for Ge, as illustrated in chapters 3 and 4, many kinetic effects such as severe GeO desorption, GeO$_2$ crystallization, and nonuniform void formation occur. These effects are rarely concentrated or undetected in Si-MOS technology because the quality of thermal oxidized SiO$_2$/ Si interface is intrinsically stable. Therefore, simply copying what we have done in the Si-MOS process to Ge-MOS is not a wise way for the interface control of GeO$_2$/Ge stacks.

In this chapter, we discuss some aspects in Ge oxidation that may closely related to the interface control of GeO$_2$/Ge stacks and provide guidelines to achieve a high-quality GeO$_2$/Ge stack.

5.2 DEAL-GROVE MODEL BREAKS DOWN FOR GE OXIDATION

The growth of thin oxide films during the thermal oxidation of silicon is well described by the Deal-Grove model,[1] in which it is considered that oxide growth proceeds only via molecular oxygen diffusion to the Si/SiO$_2$

DOI: 10.1201/9781003284802-5

interface and reaction with silicon at the interface. However, whether or not the Deal-Grove model is applicable to the oxidation of Ge is still under question. And there is still no such experiment to clarify the diffusion species in Ge oxidation.

Previously, T. Sasada et al.[3] investigated the Ge oxidation kinetic on (100), (110), and (111) substrates. A parabolic relationship was obtained between the oxide thickness and time. This result implies that the oxidation includes a diffusion process, but the exact diffusion species has not been determined yet. To investigate the diffusion process during oxidation, isotope tracing experiments will be helpful. Therefore, we investigated the oxygen profile in GeO_2 for $^{18}O_2$ oxidation.

First, 7nm-thick GeO_2 was thermally grown on a HF-last Ge (100) substrate in 1atm $^{16}O_2$ at 550°C. After that, it was re-oxidized in 20atm $^{18}O_2$ ambient at 500°C to achieve a 10nm-thick GeO_2 film. The sample structure used in this experiment is schematically shown in Figure 5.1(a).

Figure 5.1(b) shows high-resolution Rutherford backscattering spectrometry (HR-RBS) of the depth profile of ^{18}O. In Figure 5.1(b), the ^{18}O is not localized at the Ge/GeO_2 interface but spread widely in the GeO_2 film. This is quite different from the Deal-Grove model. In the Deal-Grove model, since the oxide growth is attributed to the molecular oxygen diffusion and the interfacial reaction between O_2 and substrate, so the newly formed interfacial oxide should be concentrated with the newly in-diffused oxygen, and the non-interfacial layer should contain less newly in-diffused oxygen. This is not consistent with the HR-RBS result. Therefore, our result suggests that the Deal-Grove model is not sufficient to describe oxidation of Ge/GeO_2 systems. Something other than O_2 diffusion should be taken into consideration: e.g., the oxygen atoms may diffuse by oxygen atom exchange with the existing GeO_2 film, or GeO formed at the Ge/GeO_2 interface may diffuse back to the GeO_2 film. In any case, however, the oxygen kinetics should be studied in more detail to further control Ge/GeO_2 gate stacks.

5.3 ACTIVE OXIDATION IN GEO₂/GE STACKS

5.3.1 Definition of Active Oxidation and Passive Oxidation

The reaction of oxygen with surfaces of semiconductors has attracted considerable interest, primarily due to its crucial role in the processing of microelectronic devices. For silicon, at sufficiently elevated temperatures, exposure of Si surfaces to O_2 or O will lead to the formation of volatile SiO.

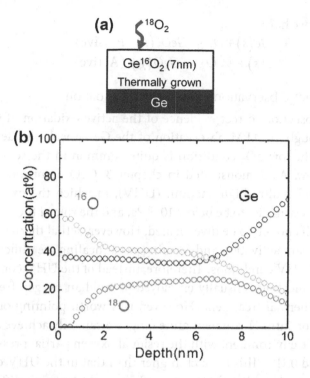

FIGURE 5.1   (a) Sample structure of Ge/GeO₂ (~10nm) stacks for ¹⁸O₂ oxidation.
(b) Depth profiles of both ¹⁸O and ¹⁶O in GeO₂ after ¹⁸O₂ oxidation by HRBS
measurement. The O profiles at both interfaces may include an artifact due to
uniformity and/or roughness, but note the relatively flat distribution of ¹⁸O in
the GeO₂ film.

This is called active oxidation, as opposed to passive oxidation, in which a
stable SiO_2 film is formed.[4, 5]

For Ge, unlike SiO_2/Si, in which the desorption of SiO is limited to only
several mono-layers, the desorption of GeO from GeO₂/Ge occurs rigorously, even in very thick cases. Therefore, the oxidation behavior of Ge
should be studied by taking into consideration GeO desorption.

Before discussing the oxidation of Ge, we have to clarify the definition
of active and passive oxidation, respectively. In this study, active oxidation refers to oxidation with the consumption of semiconductor atoms,
and passive oxidation is oxidation without the consumption of semiconductor atoms.

For Ge, we have

$$Ge(s) + O_2 \rightarrow GeO_2(s) \text{ (Passive)} \qquad (5-1)$$
$$Ge(s) + \tfrac{1}{2}O_2 \rightarrow GeO(g) \text{ (Active)} \qquad (5-2)$$

5.3.2 Direct Observation of Ge Active Oxidation

In this subsection, direct evidence of the active oxidation of Ge will be given through an AFM observation of the Ge annealed under low pO$_2$. Actually, the low pO$_2$ condition is quite common in the semiconductor process flow. As demonstrated in chapter 3, GeO desorption is mainly studied under ultra-high vacuum (UHV), in which the oxygen partial pressure is estimated to be below 10^{-10}Pa, and the role of the residual oxygen is small enough to be investigated. However, actual thermal processes such as dopant activation and forming gas annealing are difficult to carry out under UHV conditions. Therefore, instead of the UHV conditions, N$_2$ ambient annealing is usually employed to build an oxygen-free environment for thermal treatment. However, it is worth pointing out that, for example, for 1atm N$_2$ ambient annealing, we can only achieve a relatively oxygen-free environment with the residual oxygen partial pressure generally around 0.1Pa. This is much higher than that in the UHV conditions. Thus, we consider that the oxygen effect might be quite different from those in the UHV cases. Although oxidation or oxygen adsorption on a semiconductor surface has been widely investigated,[6–10] direct observation results are still very rare.

To investigate oxidation on a Ge substrate, stable oxide fins (some were 110nm sputtered Y$_2$O$_3$, and some were 160nm-thick spin-on glass SiO$_2$) with a size of 1.5um × 5mm × 160nm (width × length × thickness) were fabricated using a photolithography technique. The distance between the two GeO$_2$ fins was 1.5um, and the distance between the two SiO$_2$ fins was 7.5um. The line-patterned structure is schematically shown in Figure 5.2(a), and an AFM image of the initial surface is shown in Figure 5.2(b). By using this structure, we could find the level of the initial Ge surface because, no matter what treatment we used, the Ge surface underneath the SiO$_2$ capping was kept.

Figure 5.3 shows the cross-sectional profiles of the as-received sample and those annealed at 600°C oxygen partial pressure 0.1Pa for 30, 90, and 150 minutes, respectively. Since the SOG fins are thick enough, it is quite reasonable to believe that the reaction between oxygen and Ge substrate only occurs at the places without SOG covering. The Ge substrate surface

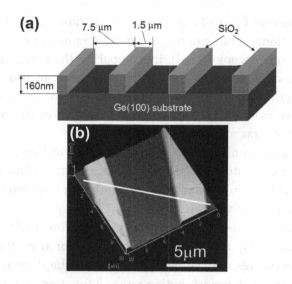

FIGURE 5.2 (a) Schematics of initial line-patterned SOG-SiO$_2$/Ge structure. (b) AFM surface image.

Source: Reproduced from S. K. Wang et al., "Desorption kinetics of GeO from GeO$_2$/Ge structure", *J. Appl. Phys.* 108, 054104 (2010), with the permission of AIP Publishing.

FIGURE 5.3 Cross-sectional profiles of as-received sample and those annealed at 600°C, oxygen partial pressure 0.1Pa for 30, 90, and 150 minutes, respectively.

Source: Reproduced from S. K. Wang et al., "Desorption kinetics of GeO from GeO$_2$/Ge structure", *J. Appl. Phys.* 108, 054104 (2010), with the permission of AIP Publishing.

without SOG fins was clearly consumed as the annealing time increased at low-oxygen partial pressure, suggesting that the Ge substrate is "etched" during the annealing treatment. (Author's note: These results give important hints for etching or reshaping Ge by controlling the active oxidation process to make novel structures.)

5.3.3 O$_2$ Pressure-Dependent Oxidation

To further investigate the oxidation kinetics of GeO$_2$/Ge, the as-prepared line-patterned samples were annealed in N$_2$/O$_2$ or O$_2$ ambient with oxygen partial pressure of 0.1~10^{7}Pa at various temperatures. Then the surface and cross-sectional profiles of the as-annealed samples were measured by atomic force microscopy (AFM). By comparing the height of the places covered with oxide fins and that of the place without fins (called consumption depth), we can determine the oxidation behavior of Ge. In other words, passive oxidation results in the minus shift of the consumption depth, and vice versa for active oxidation.

By plotting the consumption depth against annealing time at a fixed temperature, we can study the pO$_2$ dependence of the oxidation. Figure 5.4 shows the relationship of consumption depth as a function of oxidation time under various oxygen pressures at 600°C.

At low pressure (0.1~100Pa), the plots meet the linear relationship very well, suggesting only active oxidation occurs. Moreover, the consumption rate at each pressure can be extracted from the slope in Figure 5.4. Figure 5.5(a) shows the consumption rate as a function of pO$_2$.

Obviously from Figure 5.5(a), the relationship between consumption rate and pO$_2$ is non-linear; higher pO$_2$ shows a higher consumption rate. However, as we know, the probability of an O$_2$ molecule sticking to the Ge surface is proportional to pO$_2$. Therefore, the average consumption rate per O$_2$ molecule could be estimated by consumption rate/pO$_2$.

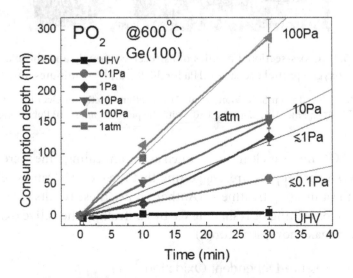

FIGURE 5.4 The relationship of consumption depth as a function of time under various oxygen pressures at 600°C.

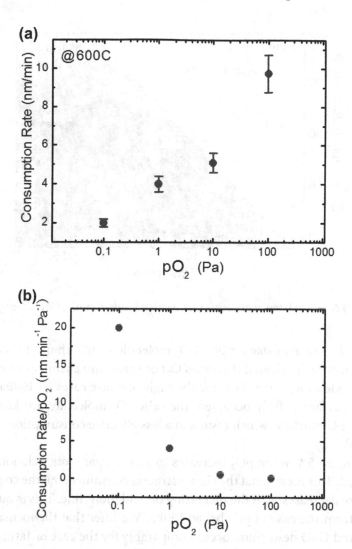

FIGURE 5.5 (a) Ge substrate consumption rate as a function of pO$_2$ at 600°C. (b) Consumption rate/pO$_2$ as a function of pO$_2$ at 600°C.

Consumption rate/pO$_2$ as a function of pressure is shown in Figure 5.5(b). In Figure 5.5(b), it is clear that an O$_2$ molecule at low pressure is more active in etching the Ge substrate than one at a higher pressure. One possible explanation is that the surface oxidation contains two steps: a physical adsorption step followed by the chemical adsorption step. The physical adsorption is a precursor state; once a site on Ge substrate is occupied, this

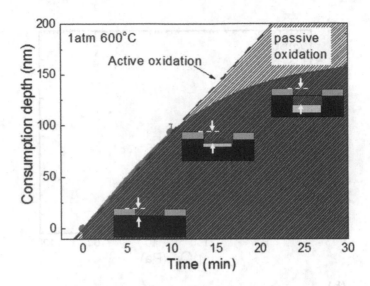

FIGURE 5.6   Coexistence of active and passive oxidation in a GeO$_2$/Ge system.

site will be not available for other O$_2$ molecule to adsorb until this physical adsorption is released (becomes O$_2$) or turns into chemical adsorption (active oxidation). Therefore, for the high-pressure cases (1~100Pa), once the Ge surface is fully occupied, the other O$_2$ molecules are kept away from the Ge surface, which results in a less effective consumption rate per molecule.

In Figure 5.4, when pO$_2$ increases to 1atm, a parabolic relationship is obtained. This means that the Ge substrate is consumed, but the consumption rate decreases with the increasing of oxidation time. This is quite different from the case of pO$_2$ below 100Pa. We infer that the formation of GeO$_2$ and GeO desorption occur comparably for the case of 1atm 600°C, as shown in Figure 5.6. The formation of GeO$_2$ becomes a barrier to active oxidation and results in a reduced consumption rate.

5.3.4 Activation Energy of Active Oxidation

In this subsection, we investigate the activation energy of active oxidation. Different from the case of GeO desorption from GeO$_2$/Ge, GeO desorption derived from surface oxidation does not include a diffusion process. Since the active oxidation is simply determined by the surface reaction, if pO$_2$ is fixed, the activation energy of Ge active oxidation can be calculated by varying the annealing temperature under low pO$_2$. Figure 5.7 shows desorption rate of active oxidation, R$_d$, as a function of temperature.

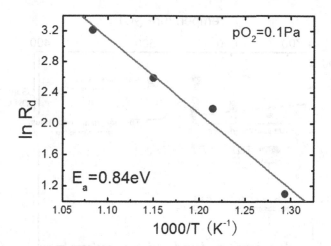

FIGURE 5.7 Desorption rate of GeO from surface active oxidation as a function of temperature. Note that the partial oxygen pressure is ~0.1Pa.

On the one hand, the desorption rate is proportional to the consumption depth of Ge substrate per minute. Thus, we have:

$$R_d = A\Delta d_{Ge}/\Delta t \qquad (5\text{--}3)$$

where Δd_{Ge} is the consumption depth of Ge substrate, and Δt is thermal treatment time. A is a constant pre-factor.

On the other hand, since active oxidation is a surface reaction limited process, we assume that the active oxidation conforms to the following Arrhenius relationship:

$$R_d = C(pO_2)k_{act} = C(pO_2)k_0 exp(-E_a/k_B T) \qquad (5\text{--}4)$$

where $C(pO_2)$ is a pre-factor that relates to the partial pressure of oxygen, k_0 is constant, k_B is Boltzmann's constant, and E_a is the surface reaction activation energy.

Therefore, by plotting R_d against $1/T$, activation energy of about 0.84 eV can be extracted from the slope.

5.4 GUIDELINES FOR INTERFACE CONTROL OF GEO$_2$/GE STACKS

5.4.1 pO$_2$-T Diagram of GeO$_2$/Ge Oxidation and GeO Desorption

Figure 5.8 shows the pO$_2$-T diagram of GeO$_2$/Ge oxidation and GeO desorption. In this diagram, the blue solid line labeled "passive/active"

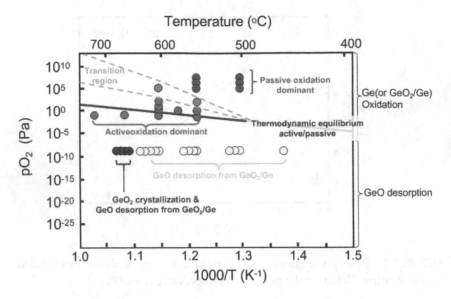

FIGURE 5.8 pO_2-T diagram of GeO_2/Ge oxidation and GeO desorption. The plots are extracted from the experimental data obtained in this study.

is the thermodynamic calculation results. Based on the HSC thermodynamic database,[11] the "passive/active" line is calculated by considering the equilibrium state between the following two reactions:

$$2GeO(g)+O_2(g) \rightarrow 2GeO_2(s) \tag{5–5}$$

$$2Ge(s)+O_2(g) \rightarrow 2GeO(g) \tag{5–6}$$

In the low-temperature and high-pressure regime, the oxidation process is governed by the passive oxidation dominantly. In this regime, there will be almost no GeO desorption in the thermal process. However, in the high-temperature and low-pressure regime, the thermal process turns out to be active oxidation dominant; in other words, the substrate is merely etched by O_2 to form GeO. Very important information in Figure 5.8 is that, between the passive dominant regime and the active dominant regime, there is a transition regime, in which both active oxidation and passive oxidation occur comparably and rigorously. This effect has never been predicted by the thermodynamic calculation. And it is also a special effect that has never been observed in the case of Si oxidation. The orange dashed lines are plotted by imagination based on the principle that the higher the temperature, the stronger both reaction fluxes are. Therefore, at high temperatures, this transition region becomes wider, while at low temperatures, it becomes narrower. It is worth pointing out that the active oxidation regime strongly "invades" across the thermodynamic equilibrium line at a temperature higher than 550°C because the activation oxidation flux is always irreversible; once Ge substrate is "etched" by O$_2$ and desorbs away in terms of GeO, the desorbed GeO will not return to the original Ge substrate. Therefore, although the GeO may be re-oxidized to GeO$_2$ somewhere other than on this substrate, the net phenomenon we observed turns out to be the etching of Ge substrate. When pO$_2$ decreases to the regime close to UHV, the extra O$_2$ is negligible. In this regime, almost no passive or active oxidation will occur on Ge substrate. However, for the GeO$_2$/Ge, GeO desorption and GeO$_2$ crystallization become dominant. Specifically, GeO desorption occurs when temperature is over 430°C in UHV, and both GeO desorption and α-quartz-like GeO$_2$ crystallization occur at a higher temperature (> 660°C).

5.4.2 pO$_2$-T Control for a GeO$_2$/Ge Stack with High Quality

In this subsection, we present some guidelines from the viewpoint of control of pressure and temperature in the GeO$_2$/Ge stack processing technique:

1. Low processing temperature (<430°C) is preferred.

   Advantage: No GeO desorption, no active oxidation, no crystallization, thin film achievable, good bulk and interface.
   Disadvantage: Chemical stoichiometry and a well-organized network are difficult to attain and require a highly active oxidant such as O* radical, O$_2$ plasma, or O$_3$.

2. If (1) cannot be satisfied, processing temperature should be less than 660°C, pO$_2$ increased, and the stack annealed at low temperature (<430°C) to repair the interface.

   Advantage: Less GeO desorption, less active oxidation, no crystallization, good bulk and interface.
   Disadvantage: Requires high pO$_2$, hard for scaling down.

3. If (1) and (2) cannot be satisfied, deposit a capping material with less oxygen diffusivity first, followed by 1atm oxidation at low temperature (500~550°C preferred), and anneal the stack at low temperature (<430°C) to repair the interface.

Advantage: Less GeO desorption, no oxidation, no crystallization, good interface.
Disadvantage: Requires capping material, not good GeO$_2$ quality, hard for scaling down.

4. If (1)–(3) cannot be satisfied, high-quality GeO$_2$/Ge stacks cannot be obtained.

5.5 A FUNDAMENTAL CONSIDERATION OF GEO$_2$/GE AND SIO$_2$/SI

Throughout this study, various kinetic effects in GeO$_2$/Ge systems have been investigated. Compared with SiO$_2$/Si, most physical properties of GeO$_2$/Ge system are quite similar. For example, Ge and Si have the same diamond-like crystal structure; they are solid semiconductors, amorphous GeO$_2$ and SiO$_2$ are constructed by GeO$_4$ tetrahedra and SiO$_4$ tetrahedra, etc.

However, in this study, some effects in GeO$_2$/Ge have been revealed to occur quite differently from SiO$_2$/Si. For example, SiO desorption is limited to a few monolayers, while GeO desorption can be observed even in GeO$_2$ thicknesses of 100nm or more. According to the reports based on diffraction techniques, the crystallization of GeO$_2$ in terms of α-quartz-like structure has been observed, while SiO$_2$ remains in a stable, amorphous state on Si, even after receiving high-temperature annealing treatment.[12] And also, GeO$_2$/Ge interface is considered to be self-passivated; it does not require the forming gas annealing process (FGA) as what has been usually done in SiO$_2$/Si.[13]

Concerning the differences between Ge and Si, Houssa et al.[13] attributed them to the viscoelastic properties of GeO$_2$ and SiO$_2$, while for the viscoelastic properties between two oxides, Micoulaut et al.[14] think that the differences lie in the different intra-tetrahedral angle and angle variations of O-Ge-O and Ge-O-Ge and O-Si-O and Si-O-Si. O-Ge-O and Ge-O-Ge have broader angle distributions than those in SiO$_2$, which make GeO$_2$ more distorted than SiO$_2$. Furthermore, on the basis of theoretical calculations, E. Artacho et al.[15] give their considerations to the difference

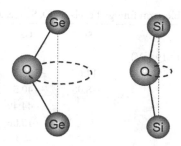

FIGURE 5.9   The spatial oxygen distribution in Ge-O-Ge and Si-O-Si.

between Ge-O-Ge and Si-O-Si bonding. Their calculations suggests that oxygen atoms in Ge-O-Ge structures are non-localized and rotate in the plane perpendicular to the axis determined by connecting the centers of two Ge atoms while, in the case of Si-O-Si, the O is almost fixed at the center between the two Si atoms. Figure 5.9 shows the schematic of the oxygen in these two cases.

E. Artacho's model is very effective to explain why GeO$_2$ is more distorted and SiO$_2$ is more rigid. However, it is not a straightforward physical explanation about the difference between GeO$_2$/Ge and SiO$_2$/Si. In both GeO$_2$/Ge and SiO$_2$/Si systems, since the O atoms are identical, the differences should originate from the difference deeply inside the Ge and Si atoms. Fundamentally, starting with their atom structures is a good way of thinking about the difference.

Si was generated from the initial big bang. Its nucleus contains 14 protons, about 14.1 neutrons on average, and 14 electrons. These 14 electrons are arranged as $1s^22s^22p^63s^23p^2$ according to quantum chemistry theory. The atomic radius of Si is about 110 ± 5pm (atomic radius in crystals).[16] The valance electrons are $3s^2$ and $3p^2$. Ge is generated from supernova explosions. Its nucleus contains 32 protons, about 40.6 neutrons on average, and 32 electrons. These 32 electrons are arranged as $1s^22s^22p^63s^23p^6\boldsymbol{3d^{10}}4s^24p^2$. The atomic radius of Ge is about 125 ± 5pm.[16] The valance electrons are $\boldsymbol{4s^2}$ and $4p^2$. Due to their similar atomic structure, Si and Ge are arranged in the same column of elements in the periodic table. However, compared with Si, the energy split for Ge valence electrons is larger, and the Ge 4s electron shows a lower energy level than Si 3s, as shown in Table 5.1.

Compared with Si, the existence of fully occupied $3d^{10}$ electrons may be the biggest difference. By solving the Schrödinger equation of Ge electrons, we know that these ten 3d electrons are statistically arranged in

TABLE 5.1   Electron Energy Levels for O, Si, Ge, and Sn[17]

Orbital	O	Si	Ge	Sn
2p	−17.19 (eV)			
3s		−14.69	−195.7	−859.9
3p		−8.08	−140.5	−740.4
3d			−44.49	−521.5
4s			−15.06	−150.0
4p			−7.82	−108.0
4d				−37.26
5s				−12.96
5d				−7.21

five orbitals with a primary quantum number n = 3. Ge and Si are quite close in atomic radii (Si: 110 pm, Ge: 125 pm), but the electron numbers are quite different (Si: 14, Ge: 32). Therefore, the 3d electrons are actually highly crowded. In this case, if we take the coulomb repulsion among the 3d electrons (strong correlation) in solving the Hamiltonian, the 3d shell will be bigger. Therefore, the overlap between the 4s cloud and 3d cloud becomes larger. This does not occur in the 3s and 3p electron case. In other words, 4s electrons have a larger probability of feeling the coulomb attraction from the nucleus. For 4p electrons, however, since their wave functions are mainly localized in the orbital away from the nucleus, the overlapping between the 4p cloud and 3d cloud is negligible. As schematically shown in Figure 5.10, due to the self-repulsion of electrons within the 3d shell, the energy difference between the 4s electron and the 4p electron is enlarged. Meanwhile, 4p electrons are less affected by the 3d electron because of their relatively larger distance to the nucleus; it is reasonable to assume that the energy level of 4p electrons is almost unchanged. So the self-repulsion of 3d electrons results in a downward shift of the 4s level. We believe the existence of 3d electrons is the fundamental reason Ge behaves more bivalently than Si, and Ge 4s electrons are more stable than Si 3s. Therefore, the s electrons of Ge are more inert than those of Si. In other words, 4s electrons in Ge have a greater probability of retaining their electrons in the reaction. This discussion reveals the intrinsic difference between Ge and Si in reaction.

Since we have already discussed the difference between Si and Ge, next part will discuss the differences in their oxides. From the viewpoint of

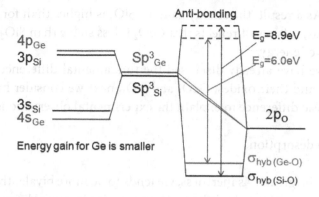

FIGURE 5.10   Schematic of Ge-O and Si-O energy diagram. After the hybridization, the energy gain for SiO_2 is higher than that of GeO_2. This means that SiO_2 has a larger possibility of sp³ hybridization than GeO_2. In other words, this result reflects that GeO_2 is less stable than SiO_2 from the viewpoint of energy.

structure, both of them are composed of SiO_4 or GeO_4 tetrahedra units. Ge's larger atomic radius allows more accessible positions to form bonds with O; therefore, GeO_2 behaves more flexibly than SiO_2.

From the viewpoint of energy, for SiO_2 and GeO_2, the valence electrons of Si or Ge form sp³ hybridization with O 2p. A simple way to compare the hybridization stability of SiO_2 and GeO_2 is to compare the energy gain between the states before and after the hybridization. Generally, a higher energy gain means a greater possibility of forming the hybridized states.

By taking account of the data in Table 5.1, we can calculate the energy gain for GeO_2 and SiO_2 hybridization. Consider the simplest case: the average energy of an electron after sp³ hybridization is:

$$E_{hyb} = \frac{1}{4}\left(E_s + 3E_p\right) \tag{5-7}$$

where E_{hyb} is the average energy of a valence electron after the sp³ hybridization, and E_s and E_p are the energy levels of s- and p-electrons without hybridization. Therefore, after the sp³ hybridization for Ge and Si, the average energy level for a Ge valence electron is a little higher than that of Si. And when these electrons form covalent bonds with O-2p electrons, the energy split between the bonding state and the anti-bonding state forms the energy gap for SiO_2 (8.9 eV) and GeO_2 (6.0 eV). Therefore, as shown in Figure 5.10, the energy level of the Si-O bond is lower than that of Ge-O. As a result, the energy gain for SiO_2 is higher than for GeO_2. In other words, this result reflects that GeO_2 is less stable than SiO_2 from the viewpoint of energy.

Since we have already discussed the fundamental difference between Ge and Si and their oxides, GeO_2 and SiO_2, next we consider how to use this intrinsic difference to explain the experimental observations.

1. GeO desorption.

Due to the greater 4s inertness, Ge tends to be more bivalently behaved in chemical reactions. And the GeO_4 tetrahedra in amorphous GeO_2 are less stable than SiO_4 in SiO_2. As a result, the reaction between Ge and GeO_2 becomes much intense than that between Si and SiO_2. And also, due to the higher bivalent nature of Ge, when GeO_2 is reduced by Vo, GeO is inclined to form.

2. GeO_2 crystallization.

Since Ge is larger than Si, it results in more accessible positions for O to form a bond. As a result, the Ge-O-Ge bonds have higher flexibility than those in SiO_2. Therefore, crystallization of GeO_2 proceeds with the help of GeO_4 distortion while SiO_2 does not. Moreover, from the viewpoint of energy, GeO_2 is less stable than SiO_2, so the network reconstruction for GeO_2 requires less energy than for SiO_2. Furthermore, since Ge tends to behave more bivalently than Si, it is possible that Vo is more easily generated at the GeO_2/Ge interface than at the SiO_2/Si interface, so higher-concentration Vo can be supplied by the interfacial reaction for the GeO_2/Ge case. As a result, GeO_2 crystallization occurs at a relatively lower temperature on Ge, while SiO_2 does not.

3. Coexistence of active and passive oxidation comparably.

Due to the inertness of 4s electrons in Ge, although the Ge substrate is covered by a GeO_2 layer, interfacial reaction–derived GeO desorption still occurs. The GeO desorption temperature lies within the temperature range for oxidation. Therefore, a transition region where both active and passive oxidation comparably occur exists.

5.6 SUMMARY

In this chapter, the Deal-Grove model is found to break down when explaining the oxidation of Ge. The active oxidation of Ge has been observed and studied. It is found that a transition region (at least within (600°C, 10^6) ~ (550°C, 100Pa)) exists between the active oxidation dominant regime and the passive dominant regime, in which both active oxidation and passive oxidation paths occur comparably.

A pO$_2$-T diagram that summarizes active oxidation, passive oxidation, transition region, GeO desorption, and GeO$_2$ crystallization is proposed. Towards the processing technique of Ge-MOS, some guidelines for preparing a high-quality GeO$_2$/Ge stack are proposed.

Finally, concerning the difference between GeO$_2$/Ge and SiO$_2$/Si, a fundamental consideration in the direction of the inertness of the 4s electrons is proposed. The inertness of Ge 4s electrons is attributed to the strong coulomb repulsion within the 3d shell.

REFERENCES

1. B. E. Deal and A. S. Grove, "General relationship for the thermal oxidation of silicon", *J. Appl. Phys.*, **36** (1965) 3770.
2. J. R. Engstrom, D. J. Bonser, M. M. Nelson, and T. Engel, "The reaction of atomic oxygen with Si (100) and Si (111): I. Oxide decomposition, active oxidation and the transition to passive oxidation", *Surface Science*, **256** (1991) 317.
3. T. Sasada, Y. Nagakita, M. Takenaka, and S. Takagi, "Surface orientation dependence of interface properties of GeO$_2$/Ge metal-oxide-semiconductor structures fabricated by thermal oxidation", *J. Appl. Phys.*, **106** (2009) 073716.
4. T. Engel, "The interaction of molecular and atomic oxygen with Si(100) and Si(111)", *Surf. Sci. Rep.*, **18** (1993) 91.
5. M. Suemitsu, Y. Enta, Y. Miyanishi, and N. Miyamota, "Initial oxidation of Si(100)-(2 3 1) as an autocatalytic reaction", *Phys. Rev. Lett.*, **82** (1999) 2334.
6. L. Surnev, "Oxygen adsorption on Ge (111) surface: I. Atomic clean surface", *Surf. Sci.*, **110** (1981) 439.
7. L. Surnev and M. Tikhov, "Oxygen adsorption on a Ge (100) surface: I. Clean surfaces", *Surf. Sci.*, **123** (1982) 505.
8. D. A. Hansen, and J. B. Hudson, "The adsorption kinetics of molecular oxygen and the desorption kinetics of GeO on Ge (100)", *Surf. Sci.*, **292** (1993) 17.
9. D. A. Hansen, and J. B. Hudson, "Oxygen scattering and initial chemisorption probability on Ge (100)", *Surf. Sci.*, **254** (1991) 222.
10. K. Nagashio, C. H. Lee, T. Nishimura, K. Kita, and A. Toriumi, "Thermodynamics and kinetics for suppression of GeO desorption by high pressure oxidation of Ge", *Mater. Res. Soc. Symp. Proc.*, **1155** (2009) C06–02.

11. HSC Chemistry® 6.0 database (OUTOKUMPU Technology).
12. A. Munkholm, S. Brennan, F. Comin, and L. Ortega, "Observation of a distributed epitaxial oxide in thermally grown SiO$_2$ on Si (001)", *Phys. Rev. Lett.*, **75** (1995) 4254.
13. M. Houssa, G. Pourtois, M. Caymax, M. Meuris, M. M. Heyns, V. V. Afanas'ev, and A. Stesmans, "Ge dangling bonds at the (100) Ge/GeO$_2$ interface and the viscoelastic properties of GeO$_2$", *Appl. Phys. Lett.*, **93** (2008) 161909.
14. M. Micoulaut, L. Cormier, and G. S. Henderson, "The structure of amorphous, crystalline and liquid GeO$_2$", *J. Phys.: Condensed Matter*, **18** (2006) R753.
15. E. Artacho, F. Yndurain, B. Pajot, R. Ramirez, C. P. Herrero, L. I. Khirunenko, K. M. Itoh, and E. E. Haller, "Interstitial oxygen in germanium and silicon", *Phys. Rev. B*, **56** (1997) 3820.
16. J. C. Slater, "Atomic radii in crystals", *J. Chem. Phys.*, **41** (1964) 3199.
17. H. Fujinaga, *Theory of Molecular Orbital*, Tokyo: Iwanami Shoten, 1979.

Printed in the United States
by Baker & Taylor Publisher Services

Printed in the United States
by Baker & Taylor Publisher Services